AF738238

Signals and Communication Technology

For further volumes:
http://www.springer.com/series/4748

E. S. Gopi

Digital Speech Processing Using Matlab

Springer

E. S. Gopi
Electronics and Communication
Engineering
National Institute of Technology
Trichy, Tamil Nadu
India

ISSN 1860-4862 ISSN 1860-4870 (electronic)
ISBN 978-81-322-1676-6 ISBN 978-81-322-1677-3 (eBook)
DOI 10.1007/978-81-322-1677-3
Springer New Delhi Heidelberg New York Dordrecht London

Library of Congress Control Number: 2013953196

© Springer India 2014
This work is subject to copyright. All rights are reserved by the Publisher, whether the whole or part of the material is concerned, specifically the rights of translation, reprinting, reuse of illustrations, recitation, broadcasting, reproduction on microfilms or in any other physical way, and transmission or information storage and retrieval, electronic adaptation, computer software, or by similar or dissimilar methodology now known or hereafter developed. Exempted from this legal reservation are brief excerpts in connection with reviews or scholarly analysis or material supplied specifically for the purpose of being entered and executed on a computer system, for exclusive use by the purchaser of the work. Duplication of this publication or parts thereof is permitted only under the provisions of the Copyright Law of the Publisher's location, in its current version, and permission for use must always be obtained from Springer. Permissions for use may be obtained through RightsLink at the Copyright Clearance Center. Violations are liable to prosecution under the respective Copyright Law.
The use of general descriptive names, registered names, trademarks, service marks, etc. in this publication does not imply, even in the absence of a specific statement, that such names are exempt from the relevant protective laws and regulations and therefore free for general use.
While the advice and information in this book are believed to be true and accurate at the date of publication, neither the authors nor the editors nor the publisher can accept any legal responsibility for any errors or omissions that may be made. The publisher makes no warranty, express or implied, with respect to the material contained herein.

Printed on acid-free paper

Springer is part of Springer Science+Business Media (www.springer.com)

Dedicated to my wife G. Viji, my son A. G. Vasig and my daughter A. G. Desna

Preface

The most of the applications of digital speech processing deal with speech or speaker pattern recognition. To understand the practical implementation of the speech or speaker recognition techniques, there is the need to understand the concepts of digital speech processing and the pattern recognition. This book aims in giving the balanced treatment of both the concepts. This book deals with speech processing concepts like speech production model, speech feature extraction, speech compression, etc., and the basic pattern recognition concepts applied to speech signals like PCA, LDA, ICA, SVM, HMM, GMM, BPN, KSOM, etc. The book is written such that it is suitable for the beginners who are doing basic research in digital speech processing. All the topics covered in this book are illustrated using Matlab in almost all the topics for better understanding.

Acknowledgments

I would like to thank Profs. S. Soundararajan (Director, NITT, Trichy), M. Chidambaram (IITM, Chennai), K. M. M. Prabhu (IITM, Chennai), P. Palanisamy, P. Somaskandan, B. Venkataramani, and S. Raghavan (NITT, Trichy) for their support. I would also like to thank those who helped directly or indirectly in bringing out this book successfully. Special thanks to my parents Mr. E. Sankara Subbu and Mrs. E. S. Meena.

Thanks

E. S. Gopi

Contents

m-Files

annforspeech.m	Artificial neural network for speech recognition.
svmforspeech.m	Support Vector Machine for speech recognition.
dosvm.m	Called by the function symforspeech.m
svmforspeechkernel.m	Support vector machine with kernel for speech recognition.
dosvmkernel.m	Called by svmforspeechkernel.m
hmmforspeech.m	Hidden Markov Model for speech.
hmmtestingforspeech.m	Testing phase for Hidden Markov Model for speech.
kmeansforspeech.m	k-means algorithm for speech.
logsum.m	Logirthm of the summation of two variables using.
gmmforspeech.m	Gaussian Mixture Model for speech.
gmmtestingforspeech.m	Testing phase for Gaussian Mixture Model for speech.
fuzzykmeansforspeech.m	Fuzzy k-means algorithm for speech.
ksomforspeech.m	Kohenen self organizing map for speech.
pcaforspeech.m	Principal component analysis for speech.
ldaforspeech.m	Linear discriminant analysis for speech.
kldaforspeech.m	Kernel-Linear discriminant analysis for speech.
k00.m	Inner-product kernel.
k11.m	Gaussian kernel.
k22.m	Polynomial kernel.
k33.m	Power exponential kernel.
k44.m	Hyperbolic tangent kernel.
k55.m	Cauchy kernel.
k66.m	Inverse multi-quadratic kernel.
icabasisforspeech.m	Independent Component Analysis (ICA) for speech.
gram.m	Gram-Schmidt orthogonalization procedure.
lattice.m	Realization of lattice structure.
levinsondurbin.m	Demonstration of Levinson Durbin algorithm.
endpointdetection.m	End-point detection of the speech segment.
ste.m	Computation of short term energy.
zcr.m	Computation of Zero crossing rate.

dynamictimewarping.m	Dynamic time warping.
vtlpc.m	Vocal tract Linear predictive co-efficients.
lpc1.m	Called by vtlpc.m
rvt.m	Roots of the vocal tract filter.
cepstrumcoef.m	Computating of cepstral co-efficients.
linespectralfrequency.m	Computation of line spectral co-efficients.
collectmfcc.m	Computation of Mel-frequency cepstral co-efficients.
obtainmfcc.m	Called by collectmfcc.m
dctbasisdemo.m	Demonstration of Discrete Cosine Transformation Basis.
spectrogram.m	Spectrogram for speech.
sg.m	Called by spectrogram.m
wavtrans.m	Wavelet transformation of the speech segment using filter decomposition method.
pitchestautocorr.m	Estimation of pitch frequency using auto-correlation method.
pitchfrequest.m	Estimation of pitch frequency using Homomorphic filtering.
endpointdetection.m	End-point detection of the speech segment.
Forcollect.m	Formant frequency computation of the speech segment.
remdup.m	Remove the duplicate number in the array.
forexthom.m	Formant extraction using homomorphic filtering.
collectf.m	Called by forexthom.m
nuqg.m	Non-uniform quantization for gaussian.
prediction.m	Demonstration of the prediction used in Differential Pulse Code Modulation.
celp.m	Code Exited Linear Prediction (CELP).
lagrangean.m	Demonstration of Lagrangen technique with equality constraints.
kuhntucker.m	Demonstration of Lagrangen technique with in-equality constraints.
preemp.m	Demonstration of usage of preemphasis filter to improve condition number.
spe.m, spelpc.m, spepre-emp.m, spepreemplpc.m	Called by preemp.m
spectralflatnessdemo.m	Measurement of spectral flatness of the excitation source.
spectralflatness.m, spectralflatnesserror.m	Called by spectralflatnessdemo.m

Chapter 1
Pattern Recognition for Speech Detection

Abstract The supervised pattern recognition techniques such as back-propagation neural network (BPNN), support vector machine (SVM), hidden Markov model (HMM), and Gaussian mixture model (GMM) that are used to design the classifier for speech and speaker detection are described in this chapter. The unsupervised techniques such as fuzzy k-means algorithm and Kohonen self-organizing map (KSOM) are discussed in this chapter. The dimensionality reduction techniques such as principal component analysis (PCA), linear discriminant analysis (LDA), kernel LDA, and independent component analysis (ICA) are also discussed in this chapter. The techniques described in this chapter are illustrated using the MATLAB for better understanding.

1.1 Introduction

Speech recognition involves identifying the isolated word from the corresponding speech signal. From the speech signal that corresponds to the particular word, one or more features such as linear predictive coefficients (LPC), Mel-frequency cepstral coefficients (MFCC) (refer Chap. 3) are collected and are arranged as the elements to form the vector. This is known as the feature extraction, and the corresponding vector is known as feature vector. Every element in the vector is known as the attributes. Suppose we need to design the digit classifier that classifies the word $zero$ to $nine$ (digits) from the corresponding speech signal. About large number (several hundreds) of feature vectors corresponding to the individual digits are collected from the various speakers to design the speaker-independent classifier. Fifty percent of the collected data are used for designing the classifier. This is generally known as training phase. The remaining 50 % are used for testing the classifier. The collected feature vectors belonging to the identical digit form the cluster. The number of clusters (class) in this example is ten.

E. S. Gopi, *Digital Speech Processing Using Matlab*,
Signals and Communication Technology, DOI: 10.1007/978-81-322-1677-3_1,
© Springer India 2014

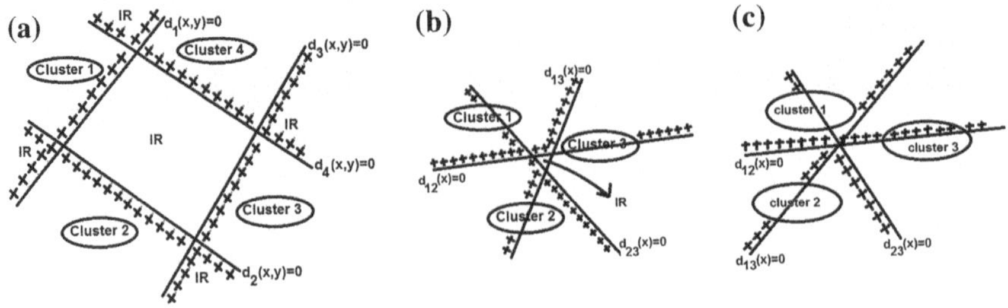

Fig. 1.1 Illustration of three types of multiclass classifier. The + in the figure indicates the region for which the corresponding boundary line equation gives positive value. **a** Type 1; **b** Type 2; **c** Type 3

The classifier identifies the separating boundary for the individual clusters. Consider the case, with the number of attributes equal to two. So the feature vectors are scattered in the 2D plane (XY plane). The coordinates on the plane are represented as (x, y). The generalized equation of the linear boundary that separates the particular cluster is given as $d(x, y) = ax + by + c = 0$. Based on the formulation of the separating boundary for the multiclass classifier, the classifier is classified as Type 1, Type 2, Type 3 (refer Fig. 1.1) as described below.

1. Type 1
   The boundary line that separates the cluster i with other feature vectors is represented as d_i. The number of decision boundary lines is equal to the number of clusters N.
   $d_i(x, y) > 0$ if $(x, y) \in$ cluster i
   $d_i(x, y) < 0$ otherwise
2. Type 2
   The boundary line that separates the cluster i with the cluster j is represented as d_{ij}. The number of decision boundary lines is $\frac{N(N-1)}{2}$, where N is the number of clusters.
   $d_{ij}(x, y) > 0$ if $(x, y) \in$ cluster i and $\forall j$
3. Type 3
   For the typical example, let us assume that Type 1 classifier is available. That is, $d_i(x, y)$ and $d_j(x, y)$ are obtained for the clusters i and j, respectively. Also note that the points on the boundary line that separate the clusters i and j satisfy $d_i(x, y) = d_j(x, y)$. Thus, from the Type 1 classifier, Type 3 classifier is obtained as follows:
   $d_{ij}(x, y) = (d_i(x, y) - d_j(x, y)) > 0$ if $(x, y) \in$ cluster i and $\forall j$

To identify the arbitrary point on the plane to conclude as the one belonging to the particular cluster, substitute in the corresponding boundary line equation and check for the cluster number. If the point favors for more than one cluster, that point is determined as the one belonging to the indeterminant region (IR). For the Type 1 and Type 2 classifiers, IR is present. But in case of Type 3 classifier, IR is absent.

The classifier with the linear boundary lines is called linear classifier (refer Fig. 1.1). There is no guarantee for the existence of linear classifier. In this case, the classifier with nonlinear boundary curves (nonlinear classifier) can be used. This is equivalently obtained as follows. First, the feature vectors are mapped to the higher-dimensional space (HDS), and the linear classifier is constructed in that space. For instance, 2D feature vector with coordinate (x, y) is mapped to the higher-dimensional space using the nonlinear transformation as $[x^2 \; y^2 \; xy]$. In this case, 2D feature dimensional space (FDS) is mapped to the 3D higher-dimensional space (HDS). Let the coordinate in the HDS be represented as (p, q, r). The linear equation (plane equation) obtained using the coordinates (p, q, r) can be equivalently represented as the nonlinear equation with x and y. Hence, the nonlinear curve in the FDS is equivalent to the linear plane in the HDS. Also by intuition, we understand that the better separation is achieved if the feature vectors are mapped to the HDS and linear classifier can be formulated easily in the HDS. Thus, in general, constructing the classifier involves identifying the set of decision rules with the set of nonlinear functions $g_i(x, y)$. The typical example is the back-propagation neural network (BPNN) as described in Sect. 1.2.

1.2 Back-Propagation Neural Network

Back-propagation neural network (BPNN) is mathematical model inspired from the biological behavior of the human brain. This is the nonlinear model usually used to model the nonlinear function. In pattern recognition, it is used to model the nonlinear functions used to formulate the set of decision rules to construct the classifier as described below. The BPNN model consists of input layer, one or more hidden layers, and output layer. Each layer is constructed using the disconnected neurons. Every neuron in the particular layer is connected to every neuron in the immediate next layer (refer Fig. 1.2). The number of neurons in the input layer is equal to the number of attributes, and the number of neurons in the output layer is equal to the number of nonlinear functions used to formulate the decision rules to construct the classifier. The number of neurons in the hidden layers is arbitrarily chosen.

Consider the BPNN (refer Fig. 1.2) with two neurons (number of attributes) in the input layer, three neurons in the hidden layer, and two neurons in the output layer. Let us assume that there are four clusters.

The arbitrary input vector, which is the input to the input layer, is represented as $[i_1 \; i_2]^T$. The vector obtained in the hidden layer is represented as $[h_1 \; h_2 \; h_3]^T = f_1([i_1 \; i_2]^T[P]_{2\times 3} + [U]_{1\times 3})$. Similarly, the vector obtained in the output layer is represented as $[o_1 \; o_2]^T = f_2([h_1 \; h_2 \; h_3]^T[Q]_{2\times 3} + [V]_{1\times 3})$. The nonlinear equations $g_1(x, y)$ and $g_2(x, y)$ that describe the decision rule (D_1 (say)) are as follows:

1. cluster 1: $g_1(i_1, i_2) < 0.5, g_2(i_1, i_2) < 0.5$
2. cluster 2: $g_1(i_1, i_2) < 0.5, g_2(i_1, i_2) > 0.5$
3. cluster 3: $g_1(i_1, i_2) > 0.5, g_2(i_1, i_2) < 0.5$
4. cluster 4: $g_1(i_1, i_2) > 0.5, g_2(i_1, i_2) > 0.5$

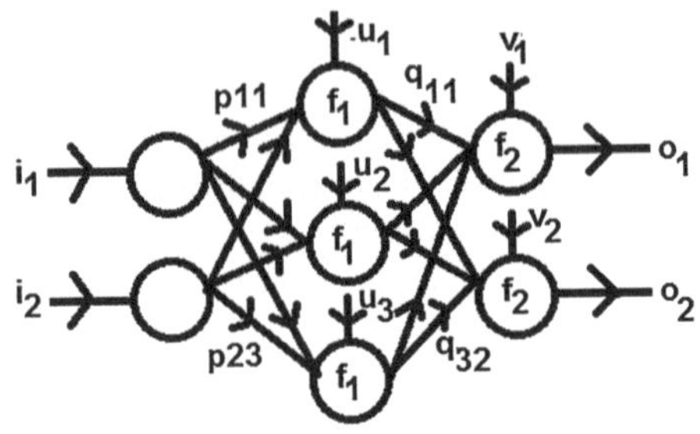

Fig. 1.2 Back-propagation neural network

It is noted that $o_1 = g_1(i_1, i_2)$ and $o_2 = g_2(i_1, i_2)$. P and Q are the weight matrices. The (i, j)th element of the weight matrix P is represented as p_{ij}. This is the connection between the ith neuron of the input layer and the jth neuron in the hidden layer. Similarly, the (i, j)th element of the weight matrix Q is represented as q_{ij}. This is the connection between the ith neuron of the hidden layer and the jth neuron in the output layer. If any of the neurons obtains the value almost zero, the neuron becomes inactive in terms of contributing to the computation of the values in the next layers. This is circumvented by using the bias vectors U and V. The ith element of the vector U is the bias value of the ith neuron of the hidden layer. Similarly, the ith element of the vector V is the bias value of the jth neuron of the output layer. The transfer functions f_1 and f_2 are used to avoid the overflow during computations. The commonly used transfer functions are the log-sigmoid (logsig) function given as $\text{logsig}(x) = \frac{1}{1+e^{-x}}$. In this example, f_1 is chosen as logsig and f_2 is chosen as linear (i.e., $f_2(x) = x$). The unknown parameters to model the BPNN (refer Fig. 1.2) are arranged as the vector as follows:

$$V = [p_{11}\ p_{12}\ p_{13}\ p_{21}\ p_{22}\ p_{23}\ q_{11}\ q_{12}\ q_{21}\ q_{22}\ q_{31}\ q_{32}\ u_1\ u_2\ u_3\ v_1\ v_2]^T.$$

To construct the BPNN that satisfies the decision rule D_1, the unknown vector V is optimized such that following target vectors (output) are obtained for the corresponding input feature vectors.

1. If the input feature vectors are from the first cluster, the target output vector is fixed as $[0.01\ 0.01]^T$.
2. If the input feature vectors are from the second cluster, the target output vector is fixed as $[0.01\ 0.99]^T$.
3. If the input feature vectors are from the third cluster, the target output vector is fixed as $[0.99\ 0.01]^T$.
4. If the input feature vectors are from the fourth cluster, the target output vector is fixed as $[0.99\ 0.99]^T$.

In general, it is not possible to obtain the typical values of the target vector for the corresponding input feature vectors. So we optimize that vector V that minimizes the sum-squared error (SSE) between the actual output vector and the corresponding target vector as described below. Let us define that the actual output vector obtained corresponding to the ith feature vector is O^i and corresponding fixed target vector is represented as R^i. Hence, we have to optimize by minimizing the sum-squared error obtained as follows:

$$\mathrm{SSE} = \sum_{n=1}^{n=N} (R^n - O^n)^T (R^n - O^n) \tag{1.1}$$

N is the number of feature vectors used for training. Obtaining the optimal values of the unknown vector V by minimizing (1.1) is known as the training phase. The feature vectors are equally collected from all the four clusters. Usually, fifty percent of the collected vectors are used for the training phase. The remaining fifty percent are used for the testing phase using the decision rule D_1. The performance of the trained BPNN is measured in terms of the SSE obtained using (1.1) and the percentage of success (POS) obtained using the trained BPNN using the testing data. Minimizing (1.1) is done using the back-propagation algorithm based on steepest descent algorithm as described below.

1.2.1 Back-Propagation Algorithm

1. Initialize the vector V randomly.
2. Obtain the output vector corresponding to the first training vector. Compute the corresponding error vector.
3. The vector V is updated using the error vector based on steepest descent algorithm as described below.
4. Obtain the output vector corresponding to the next training vector with the latest updated vector V (refer Sect. 1.2).
5. Steps 3 and 4 are repeated for all the N training vectors. This is one epoch.
6. Compute the SSE using the latest updated vector V (refer (1.1)).
7. Go to step 2, if the number of epochs is less than the predefined number of iterations or SSE is less than the predefined value.
8. Thus, the final vector V completely describes the model.

The next best vector that minimizes the squared error for the particular input feature vector $[i_1\ i_2]$ is in the direction of the negative gradient of the squared error $J = \sum_{j=1}^{j=2}(o_j - r_j)^2$. This is the steepest descent algorithm. Thus, the training update equation is as follows:

$$V(t+1) = V(t) - \frac{1}{2}\nabla J(t) \tag{1.2}$$

where t is the iteration number. The update equation for every element of the unknown vector V is obtained as follows:

$$J = \sum_{j=1}^{j=2} (r_j - o_j)^2 \tag{1.3}$$

$$\Rightarrow J = \sum_{j=1}^{j=2} (r_j - q_{1j}h_1 - q_{2j}h_2 - q_{3j}h_3 - v_j)^2 \tag{1.4}$$

$$\Rightarrow \frac{\partial J}{\partial q_{11}} = -2(r_1 - o_1)h_1 \tag{1.5}$$

$$\Rightarrow \frac{\partial J}{\partial q_{11}} = -2e_1 h_1 \tag{1.6}$$

Thus, the update equation for q_{11} is given as

$$q_{11}(t+1) = q_{11}(t) - \frac{1}{2}\frac{\partial J}{\partial q_{11}} q_{11}(t+1) = q_{11}(t) + e_1 h_1 \tag{1.7}$$

In general, the update equation for q_{mn} is given as follows:

$$q_{mn}(t+1) = q_{mn}(t) + e_n h_m \tag{1.8}$$

Similarly, the update equation for v_m is given as follows:

$$v_m(t+1) = v_m(t) + e_m \tag{1.9}$$

The update equations for the elements of the weigh matrix P are derived as follows. p_{11} is updated by obtaining by computing $\frac{\partial J}{\partial p_{11}}$

$$J = \sum_{j=1}^{j=2} (r_j - q_{1j}h_1 - q_{2j}h_2 - q_{3j}h_3 - v_j)^2 \tag{1.10}$$

$$h_1 = f_1(i_1 p_{11} + i_2 p_{21} + u_1) \tag{1.11}$$

$$\Rightarrow \frac{\partial J}{\partial p_{11}} = \frac{\partial J}{\partial h_1}\frac{\partial h_1}{\partial p_{11}} \tag{1.12}$$

$$\frac{\partial h_1}{\partial p_{11}} = f_1'(i_1 p_{11} + i_2 p_{21} + u_1) i_1 \tag{1.13}$$

$$\frac{\partial J}{\partial h_1} = -2\sum_{j=1}^{j=2} e_j q_{1j} \tag{1.14}$$

We know

$$f_1(x) = \frac{1}{1+e^{-x}} \Rightarrow e^{-x} = \frac{(1 - f_1(x))}{f_1(x)}$$

$$\Rightarrow f_1'(x) = -\frac{e^{-x}}{(1+e^{-x})^2} = -\frac{(1 - f_1(x)) f_1(x)^2}{f_1(x)} = -f_1(x)(1 - f_1(x))$$

$$
\begin{aligned}
&\Rightarrow f_1'(i_1 p_{11} + i_2 p_{21} + u_1)\\
&= -f_1(i_1 p_{11} + i_2 p_{21} + u_1)(1 - f_1(i_1 p_{11} + i_2 p_{21} + u_1))\\
&= -h_1(1 - h_1)
\end{aligned}
$$

From (1.11)–(1.14), $\frac{\partial h}{\partial p_{11}}$ is given as $-2\sum_{j=1}^{j=2} e_j q_{1j}(h_1 - 1)h_1 i_1$. Thus, the update equation for the variable p_{11} is given as follows:

$$p_{11}(t+1) = p_{11}(t) - \frac{1}{2}\frac{\partial J}{\partial p_{11}} \tag{1.15}$$

$$p_{11}(t+1) = p_{11}(t) + \sum_{j=1}^{j=2} e_j q_{1j}(h_1 - 1)h_1 i_1 \tag{1.16}$$

Thus, the update equation for p_{mn} is given as follows:

$$p_{mn}(t+1) = p_{mn}(t) + \sum_{j=1}^{j=2} e_j q_{nj}(h_n - 1)h_n i_m \tag{1.17}$$

Similarly, the update equation for u_m is given as follows:

$$u_n(t+1) = u_n(t) + \sum_{j=1}^{j=2} e_j q_{nj}(h_n - 1)h_n \tag{1.18}$$

1.2.2 ANN Illustration

1. Three isolated speech segments are collected. Ten samples for each word are used.
2. 50 % of the collected data are used as the training set, and the remaining 50 % are used as the testing set.
3. The collected speech segments are subjected to endpoint detection and dynamic time warping (refer Chap. 3).
4. The warped speech segments are modeled using LPC. The number of LPC coefficients collected from every speech segment is 182 (refer Fig. 1.3).
5. The 182 dimensional vectors are mapped to the lower-dimensional space using null-space linear discriminant analysis (N-LDA).
6. In one of the version of N-LDA (refer Sect. 1.7.6), the vectors are projected to the null space of within-class scatter matrix. The between-class scatter matrix is computed using the projected vectors. The eigenvectors of the between-class scatter matrix correspond to the significant eigenvalues for the transformation matrix.

7. The number of eigenvectors corresponding to the significant eigenvalues is 2. The projected vectors using the two eigenvectors are displayed in Fig. 1.4.
8. The ANN is formulated using 2 neurons in the input layer, 10 neurons in the hidden layer, and 3 neurons in the output layer.
9. The target vector is chosen as $[1\ 0.3\ 0.3]^T$, $[0.3\ 1\ 0.3]^T$, and $[0.3\ 0.3\ 1]^T$ for the isolated words 1, 2, and 3, respectively.
10. After 5,000 iterations (refer Fig. 1.5), sum-squared error (SSE) reaches 3.9389. The percentage of detection is given as 93.33 %.

```
%annforspeech.m
%Dimensionality reduction LDA is done before subjected to ANN
%TRAINLPC is the mat file consists of LPC of three isolated words.
%5 samples are available for each word. The NaN in the DATA is replaced with eps.
load TRAINLPC
s=[];
for i=1:1:3
    s=[s size(TRAINLPC{i},1)];
end
x=min(s);
DATA=[];
for i=1:1:3
    DATA=[DATA TRAINLPC{i}(1:1:x,:)];
end
U=isnan(DATA);
[R,S]=find(U==1);
for i=1:1:length(R)
    DATA(R(i),S(i))=eps;
end
SW=cov(DATA(:,1:1:5)',1)+cov(DATA(:,6:1:10)',1)+cov(DATA(:,11:1:15)',1);
[E1,V1]=eigs(SW,12);
for i=1:1:15
PR(:,i)=DATA(:,i)-E1*E1'*DATA(:,i);
end
M1=mean(PR(:,1:1:5)');
M2=mean(PR(:,6:1:10)');
M3=mean(PR(:,11:1:15)');
SB=cov([M1' M2' M3']',1);
[E,D]=eigs(SB,2);
%Projected vectors (P)
P=E'*DATA;
mi=min(min(P));
P=P-mi+eps;
ma=max(max(abs(P)));
P=P/ma;
H=10;
%Backpropagation neural network with 9X18X10
W1=rand(2,H);
B1=rand(H,1);
W2=rand(H,3);
B2=rand(3,1);
TARGET=[repmat([1 0 0]',1,5)  repmat([0 1 0]',1,5) repmat([0 0 1]',1,5) ];
 TARGET=TARGET*0.7+0.3;
SSE=[]
m=max(max(abs(P)));
P=P/m;
for epoch=1:1:500
for i=1:1:15
```

```
input=P(:,i);
target=TARGET(:,i);
temp1=input'*W1+B1';
temp2=1./(1+exp(-temp1));
temp3=temp2*W2+B2';
error=target'-temp3;
for i=1:1:H
    for j=1:1:3
W2(i,j)=W2(i,j)+LP*error(j)*temp2(i);
    end
end
for i=1:1:3
B2(i)=B2(i)+LP*error(i);
end
for i=1:1:2
    for j=1:1:H
        for k=1:1:3
W1(i,j)=W1(i,j)+LP*error(k)*W2(j,k)*temp2(j)*(temp2(j)-1)*input(i);
    end
    end
end
for i=1:1:H
    for k=1:1:3
B1(i)=B1(i)+LP*error(k)*W2(i,k)*temp2(i)*(temp2(i)-1);
    end
end
        end
        RES1=W1'*P+repmat(B1,1,15);
        RES2=1./(1+exp(-RES1));
        RES3=W2'*RES2+repmat(B2,1,15);
        SSE=[SSE sum(sum((RES3-TARGET).^2))]
        end
        figure
        plot(SSE)
        %Testing
        load TESTLPC
        DATA=[];
        for i=1:1:3
            DATA=[DATA TESTLPC{i}(1:1:x,:)];
        end
        U=isnan(DATA);
        [R,S]=find(U==1);
        for i=1:1:length(R)
            DATA(R(i),S(i))=eps;
        end
        P=E'*DATA;
        P=P-mi+eps;
        P=P/ma;
        RES1=W1'*P+repmat(B1,1,15);
        RES2=1./(1+exp(-RES1));
        RES3=W2'*RES2+repmat(B2,1,15);
        [P,Q]=max(RES3);
        REF=[ones(1,5) ones(1,5)*2 ones(1,5)*3];
        POS=(length(find((REF-Q)==0))/length(REF))*100;
```

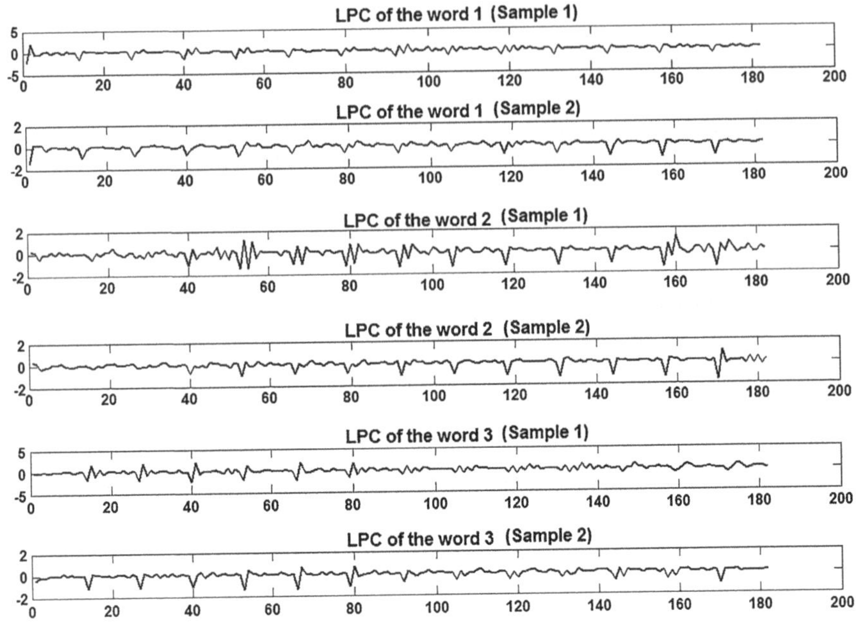

Fig. 1.3 LPC features extracted from the speech signal

1.3 Support Vector Machine

Support vector machine (SVM) comes under the category of Type 1 classifier. SVM identifies the decision boundary (refer Fig. 1.6) between the particular cluster and others. Let the linear decision boundary (line in the 2D space and the plane in HDS) that separates the cluster 1 and cluster 2 be represented as $W^T X + b = 0$, where X is the feature column vector with size $n \times 1$, W is the column vector with size $n \times 1$, and b is the scalar constant. Consider another two planes represented as $W^T X + b = 1$ and $W^T X + b = -1$ that are parallel to $W^T X + b = 0$ (refer Fig. 1.6). The goal is to obtain the optimal values for the W and b such that the following decision rule (D_2 (say)) is satisfied.

- The region $W^T X + b > 1$ belongs to the cluster 1.
- The region $W^T X + b < -1$ belongs to the cluster 2.
- The shortest Euclidean distance between the two planes $W^T X + b = 1$ and $W^T X + b = -1$ is maximized.
- Note that $W^T X + b = 0$ lies in the middle of the two planes $W^T X + b = 1$ and $W^T X + b = -1$.

The distance between the two planes $W^T X+b = -1$ and $W^T X+b = 1$ is obtained as follows. This is obtained as the difference between the minimal (perpendicular) distances computed from the origin to any arbitrary point X_1 on the plane $W^T X+b = 1$ (refer Fig. 1.7) and X_2 on the plane $W^T X+b = -1$. The distance between the origins

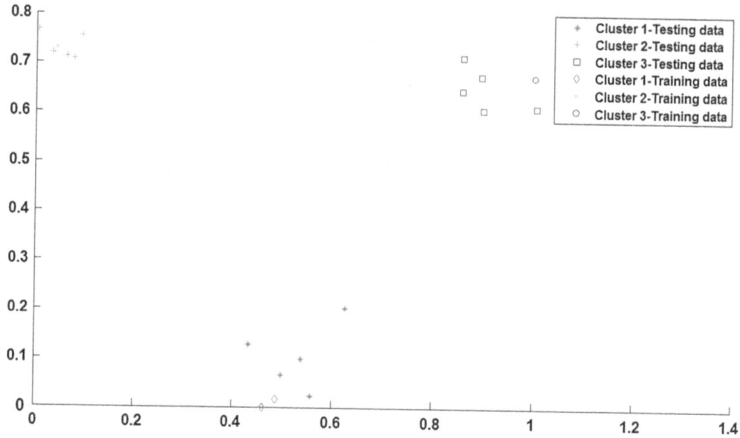

Fig. 1.4 2D scatter plot after lower-dimensional projection using N-LDA

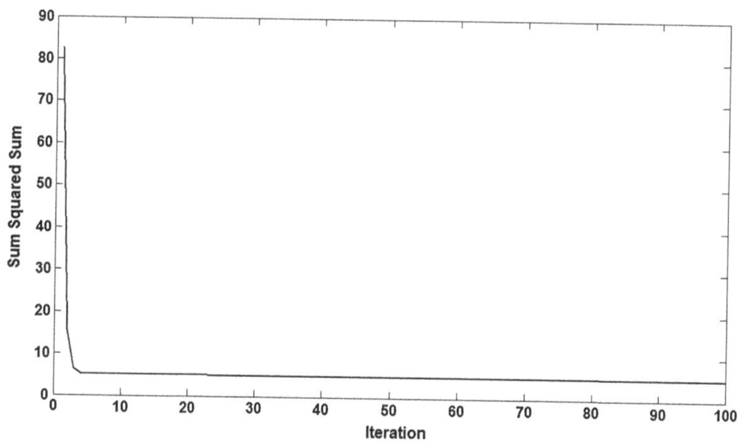

Fig. 1.5 Convergence graph for back-propagation neural network

with the arbitrary point (unknown) X_1 on the plane is $W^T X + b = 1$ and is computed as $X_1^T X_1$. The unknown arbitrary point X_1 on the plane $W^T X + b = 1$ that minimizes $x_1^T x_1$ is obtained using the Lagrangian minimization problem as described below.

Minimize $X_1^T X_1$ such that $W^T X_1 + b = 1$. The Lagrangian equation is formulated as follows:

$$J(X_1, \lambda) = X_1^T X_1 + \lambda(W^T X_1 + b - 1) \tag{1.19}$$

differentiating J with respect to X_1 and equate to zero.

$$\frac{dJ}{dX_1} = 2X_1 + \lambda(W) = 0 \tag{1.20}$$

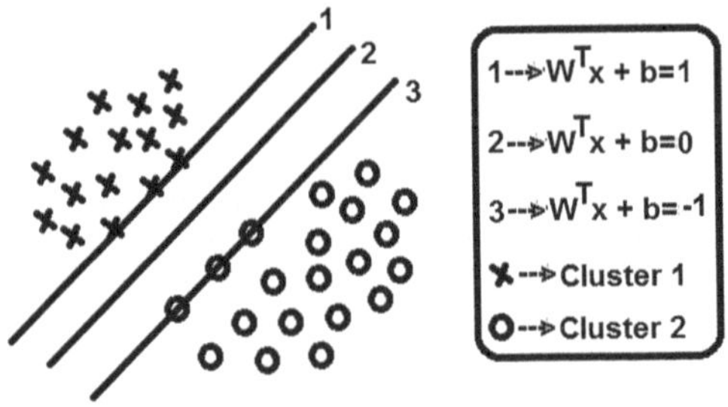

Fig. 1.6 Illustrations of support vector machine

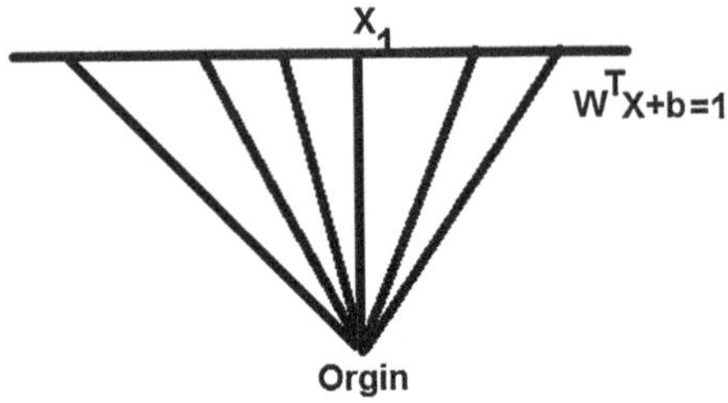

Fig. 1.7 Illustration of computing the point X_1 on the plane $W^T X + b = 1$, whose distance from the origin is minimum.

$$\Rightarrow X_1 = -\frac{1}{2}\lambda(W) \tag{1.21}$$

Substituting (1.21) in $W^T X_1 + b - 1 = 0$ and solving for λ, we get the following.

$$-W^T \frac{1}{2}\lambda(W) + b - 1 = 0 \Rightarrow -\frac{1}{2}\lambda = \frac{1-b}{W^T W} \tag{1.22}$$

Thus, the minimal distance is given as $X_1^T X_1 = \frac{1}{4}\lambda^2 W^T W = \frac{(1-b)^2}{(W^T W)^2}(W^T W) = \frac{(1-b)^2}{(W^T W)}$.

Similarly, the minimal distance between the arbitrary point X_2 on the plane $W^T X + b + 1 = 0$ with the origin is obtained as $\frac{(1+b)^2}{(W^T W)}$. Thus, the shortest distance between the two planes is given as $\frac{(1+b)^2}{(W^T W)} - \frac{1}{4}\frac{(1-b)^2}{(W^T W)} = \frac{2b}{(W^T W)}$. Our objective is to optimize the vector W and the scalar b such that the distance $\frac{2b}{(W^T W)}$ is maximized. This is equivalently achieved by minimizing $\frac{(W^T W)}{2}$ that separates the two clusters (refer Fig. 1.6). Let y_k be the index variable associated with the kth vector X_k in the training set and is represented as follows:

$$y_k = 1 \quad \text{if } k \in \text{cluster 1}$$
$$y_k = -1 \quad \text{if } k \in \text{cluster 2}$$

From the decision rule D_2, we obtain the inequality constraints using the index variables of all the training vectors as $(W^T X_k + b)y_k \geq 0$. Thus, the optimization problem for solving SVM is described as follows. Minimize $\frac{(W^T W)}{2}$ such that constraint $(W^T X_k + b)y_k \geq 1$ is satisfied. The obtained optimization problem is solved using the Lagrangian techniques with inequality constraints (refer Appendix A). The Lagrangian equation is formulated as follows:

$$J(W, b, \lambda_k) = \frac{(W^T W)}{2} - \sum_{k=1}^{k=N} \lambda_k((W^T X_k + b)y_k - 1) \tag{1.23}$$

where N is the number of training data used for optimization. Note the negative sign in the constraint part (refer Appendix A). To optimize W and b, differentiate (1.23) with respect to W, b, and λ_k and equate to zero as given below.

$$\frac{\partial J}{\partial W} = W - \sum_{k=1}^{k=N} \lambda_k X_k y_k = 0 \tag{1.24}$$

$$\Rightarrow W = \sum_{k=1}^{k=N} \lambda_k X_k y_k \tag{1.25}$$

$$\frac{\partial J}{\partial b} = \sum_{k=1}^{k=N} \lambda_k y_k = 0 \tag{1.26}$$

$$\frac{\partial J}{\partial y_k} = (W^T X_k + b)y_k - 1 = 0 \tag{1.27}$$

$$\Rightarrow (W^T X_k + b)y_k = 1 \tag{1.28}$$

Solving the equations satisfying (1.25)–(1.28) gives the optimal solution for unknown vector W and constant b. But according to Kuhn–Tucker conditions (refer Appendix A), the following equations are to be satisfied to minimize the objective function $\frac{(W^T W)}{2}$

1. $\lambda_k \geq 0$.
2. $\lambda_k((W^T X_k + b)y_k - 1) = 0$.
3. It is noted if $\lambda_k = 0$, the constraint $(W^T X_k + b)y_k - 1$ need not be equal to zero. But if $\lambda_k > 0$, the constraint $(W^T X_k + b)y_k - 1 = 0$.

In practice, it is not possible to obtain all the λ_k values positive, and hence, the training vectors X_k corresponding to the non-negative λ_k are chosen and are used to obtain the unknown vector W using (1.25). The corresponding training vectors are known as support vectors, and hence, this technique is known as support vector machine (SVM).

1.3.1 Dual Problem to Solve (1.25)–(1.28)

The dual problem is obtained by back substituting (1.25) in (1.23) to obtain the equation as the function of λ as follows:

$$J(W, b, \lambda_k) = \frac{(W^T W)}{2} - \sum_{k=1}^{k=N} \lambda_k((W^T X_k + b)y_k - 1)$$

After substituting (1.25), we get

$$L(\lambda_k, b) = \frac{1}{2}\left(\sum_{m=1}^{m=N} \lambda_m X_m y_m\right)^T \left(\sum_{n=1}^{n=N} \lambda_n X_n y_n\right) - \sum_{k=1}^{k=N} \lambda_k \left(\left(\left(\sum_{l=1}^{l=N} \lambda_l X_l y_l\right)^T X_k + b\right) y_k - 1\right)$$

$$\Rightarrow L(\lambda_k, b) = \frac{1}{2}\sum_{m=1}^{m=N}\sum_{n=1}^{n=N} \lambda_m \lambda_n y_m y_n X_m^T X_n - \sum_{k=1}^{k=N}\sum_{l=1}^{l=N} \lambda_k \lambda_l y_k y_l X_k^T X_l - b\sum_{k=1}^{k=N} \lambda_k y_k + \sum_{k=1}^{k=N} \lambda_k$$

$$\Rightarrow L(\lambda_k, b) = -\frac{1}{2}\sum_{m=1}^{m=N}\sum_{n=1}^{n=N} \lambda_m \lambda_n y_m y_n X_m^T X_n - b\sum_{k=1}^{k=N} \lambda_k y_k + \sum_{k=1}^{k=N} \lambda_k$$

After substituting (1.26), we get,

$$\Rightarrow L(\lambda_k) = -\frac{1}{2}\sum_{m=1}^{m=N}\sum_{n=1}^{n=N} \lambda_m \lambda_n y_m y_n X_m^T X_n + \sum_{k=1}^{k=N} \lambda_k$$

Note that the obtained dual problem $L(\lambda_k)$ is the unconstrained maximization problem. The solution is obtained by differentiating $L(\lambda_k)$ with respect to λ_k and equate to zero as described below.

$$L(\lambda_k) = -\frac{1}{2}\sum_{l=1}^{l=N}\sum_{k=1}^{k=N} \lambda_l \lambda_k y_l y_k X_l^T X_k + \sum_{k=1}^{k=N} \lambda_k \quad (1.29)$$

$$\frac{dL}{d\lambda_k} = -\sum_{l=1}^{l=N} y_l \lambda_l X_l^T X_k - \lambda_k X_k^T X_k + 1 = 0 \quad (1.30)$$

Thus, λ_k's are obtained using the following equations:

1. Set of N equations obtained using (1.30) for $k = 1, \ldots, n$.
2. $\sum_{k=1}^{k=N} \lambda_k y_k = 0$

Finally, the unknown W and b are obtained as follows:

1. Using the vectors corresponding to $\lambda_k \geq 0$, the vector W is computed using (1.25).

2. From the Kuhn–Tucker condition described in Sect. 1.3, the scalar constant b is obtained by solving the constraint $(W^T X_k + b)y_k = 1$ for k corresponding to the positive λ_k.

1.3.2 "Kernel-Trick" for Nonlinear Separation in SVM

It is usually difficult to identify the linear separation plane as discussed in Sect. 1.3 in the feature dimensional space. Hence, the vectors are mapped to the HDS for better separation. By intuition, we understand that it is comparatively easier to identify the separating plane in the HDS. Give the arbitrary vector X_k in the feature dimensional space, the corresponding transformed vector in the HDS is represented as $\phi(X_k)$. The dual SVM problem formulated using the HDVs is described using the following equations:

$$\lambda_k = \frac{1}{2}\sum_{l=1}^{l=N} y_l \lambda_l \phi(X_l)^T \phi(X_k) \tag{1.31}$$

for $k = 1, \ldots, N$.

$$\sum_{k=1}^{k=N} \lambda_k y_k = 0 \tag{1.32}$$

Similar to the discussion in Sect. 1.3.1, the optimal values for the W^ϕ and b^ϕ in the HDS are obtained as follows:

1. Using the vectors corresponding to $\lambda_k \geq 0$, the vector W^ϕ is computed using $W^\phi = \sum_{k=1}^{k=N} \lambda_k \phi(X_k) y_k$
2. From the Kuhn–Tucker condition described in Sect. 1.3, the scalar constant b is obtained by solving the constraint $(W_\phi^T \phi(X_k) + b_\phi)y_k = 1$ for k corresponding to the positive λ_k.

Identifying the transfer function ϕ is the difficult task. But from the Eq. (1.31), we understand that there is no need for explicit representation of the transfer function ϕ. The kernel function is defined to compute the inner product of two vectors in the HDS (i.e., $\text{kernel}(X_l, X_k) = \phi(X_l)^T \phi(X_k)$. Using the kernel function, (1.31) is rewritten as follows:

$$\lambda_k = \frac{1}{2}\sum_{l=1}^{l=N} y_l \lambda_l \text{kernel}(X_l, X_k) \tag{1.33}$$

for $k = 1 \ldots N$.

It is noted that W_ϕ vector in the HDS is not explicitly calculated. Instead, to test whether the arbitrary vector in the feature dimensional space X_{test} belongs to cluster 1 or cluster 2, we need to check the value of $W_\phi^T \phi(X_{\text{test}}) + b_\phi$ (refer the decision rule D_2). This is in further computed as follows:

$$W_\phi^T \phi(X_{\text{test}}) + b^\phi = \sum_{k=1}^{k=N} \lambda_k \phi(X_k) y_k \phi(X_{\text{test}}) + b_\phi$$

$$\Rightarrow W_\phi^T \phi(X_{\text{test}}) + b^\phi = \sum_{k=1}^{k=N} \lambda_k y_k \text{kernel}(X_k, X_{\text{test}}) + b_\phi$$

Also, it is noted that the value of b_ϕ is computed by solving $(W_\phi^T \phi(X_k) + b_\phi) y_k = 1$ for k corresponding to the positive λ_k using the kernel function, which is described as follows:

$$(W_\phi^T \phi(X_k) + b^\phi) y_k = 1$$

$$\Rightarrow \left(\left(\sum_{l=1}^{l=N} \lambda_l \phi(X_l) y_l \right)^T \phi(X_k) + b_\phi \right) y_k = 1$$

$$\Rightarrow \left(\sum_{l=1}^{l=N} \lambda_l \phi(X_l)^T \phi(X_k y_l) + b_\phi \right) y_k = 1$$

$$\Rightarrow \left(\sum_{l=1}^{l=N} \lambda_l \text{kernel}(X_l, X_K) y_l + b_\phi \right) y_k = 1$$

Thus, the SVM is performed in HDS without actually mapping the vectors to the HDS. This is known as "kernel-trick." Let $K = [\phi(X_1)\ \phi(X_2)\ \phi(X_3) \ldots \phi(X_N)]$. The Gram matrix G is obtained as $K^T K$ (i.e., the (i, j)th element of the Gram matrix G is computed as kernel(X_i, X_j)). It is noted that the valid kernel function kernel(X_i, X_j) is represented as $\phi(X_i)^T \phi(X_j)$. Hence, the Gram matrix is the diagonalizable matrix with non-negative diagonal elements. This is Mercer's condition to test the validity of the kernel function. The most commonly used kernel functions are listed in Table 1.1. It is also noted that the hyperbolic kernel function does not satisfy the Mercer's condition. The other kernel functions can be constructed using the existing kernel functions using the properties listed in Table 1.2.

1.3.3 Illustration for Support Vector Machine

LPC coefficients for three different isolated words are collected. Ten data for every isolated word are collected. 50 % of the collected data are used for training, and the remaining 50 % data are used for testing. They are subjected to N-LDA (refer Sect. 1.7.6) for dimensionality reduction. The projected vectors in the lower-dimensional space (2D) are subjected to support vector machine without kernel-trick, and the separating lines are obtained in Figs. 1.8, 1.9 and 1.10. It is noted that line separates the training data. The percentage of success in correct classification is obtained

Table 1.1 List of commonly used kernel functions

Name	$\text{kernel}(X_i, X_j)$
Inner product kernel	$(X_i^T X_j)$
Gaussian	$\exp^{\frac{-\|X_i - X_j\|^2}{c1}}$
Polynomial	$(X_i^T X_j + c2)^{c3}$
Power exponential	$(\exp^{\frac{-\|X_i - X_j\|^2}{c4^2}})^5$
Hyperbolic tangent	$\tanh(c6(X_i^T X_j) + c7)$
Cauchy	$\frac{1}{1+\frac{\|X_i - X_j\|^2}{c8}}$
Inverse multiquadric	$\frac{1}{\sqrt{\|X_i - X_j\|^2 + c9^2}}$

Table 1.2 Constructing the new kernel functions using the existing kernel functions $\text{kernel}_1(X_i, X_j)$ and $\text{kernel}_2(X_i, X_j)$

New kernel function	Remark
$c\ \text{kernel}_1(X_i, X_j)$	c is constant
$q\ \text{kernel}_1(X_i, X_j)$	q is the polynomial with non-negative coefficients
$\exp(\text{kernel}_1(X_i, X_j))$	exp is the exponential function
$\text{kernel}_1(X_i, X_j) + \text{kernel}_2(X_i, X_j)$	–

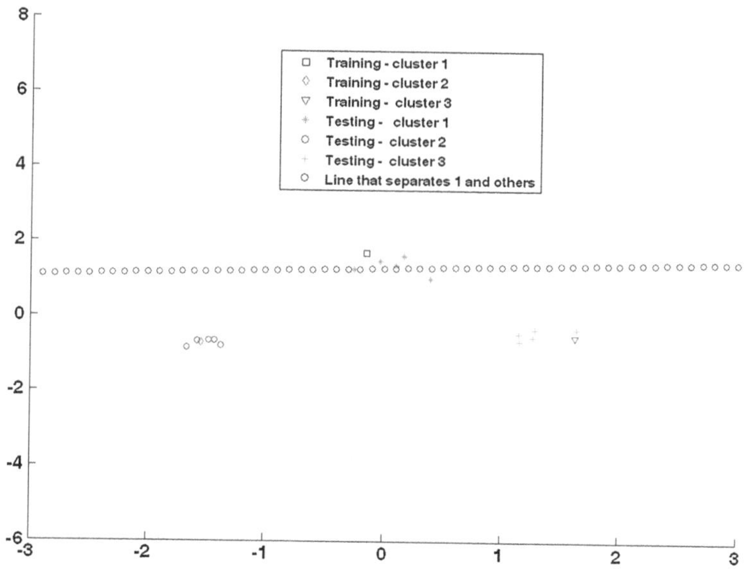

Fig. 1.8 Illustration of SVM—Part 1

as 86.67 % (13 out of 15). The experiment is repeated with Gaussian kernel, and the percentage of success in correct classification is obtained as 100 % (15 out of 15).

```
%svmforspeech.m
%Dimensionality reduction using LDA before subjected to SVM
load TRAINLPC
%TRAINLPC is the mat file consists of LPC of three isolated words.
%5 samples are available for each word. The NaN in the DATA is
%replaced with eps.
s=[];
for i=1:1:3
    s=[s size(TRAINLPC{i},1)];
end
x=min(s);
DATA=[];
for i=1:1:3
    DATA=[DATA TRAINLPC{i}(1:1:x,:)];
end
U=isnan(DATA);
[R,S]=find(U==1);
for i=1:1:length(R)
    DATA(R(i),S(i))=eps;
end
SW=cov(DATA(:,1:1:5)',1)+cov(DATA(:,6:1:10)',1)+cov(DATA(:,11:1:15)',1);
[E1,V1]=eigs(SW,12);
for i=1:1:15
PR(:,i)=DATA(:,i)-E1*E1'*DATA(:,i);
end
M1=mean(PR(:,1:1:5)');
M2=mean(PR(:,6:1:10)');
M3=mean(PR(:,11:1:15)');
SB=cov([M1' M2' M3']',1);
[E,D]=eigs(SB,2);
%Projected vectors (P)
P=E'*DATA;
M=P'*P;
CLUSTER1=1:1:5;
Y=[ones(1,5) ones(1,10)*(-1)];
POS1=dosvm(M,Y,P,x,E,CLUSTER1);
CLUSTER2=6:1:10;
Y=[ones(1,5)*(-1) ones(1,5) ones(1,5)*(-1)];
POS2=dosvm(M,Y,P,x,E,CLUSTER2);
CLUSTER3=11:1:15;
Y=[ones(1,10)*(-1) ones(1,5)*(1)];
POS3=dosvm(M,Y,P,x,E,CLUSTER3);
```

```
%dosvm.m
function [POS]=dosvm(M,Y,P,x,E,CLUSTER)
Y1=repmat(-1*Y,15,1);
```

```
M1=M.*Y1;
M2=[M1;Y];
N=[diag(M1)-1 ;0];
lambda=pinv(M2)*N;
[POS1,POS2]=find(lambda>=0);
W=0;
for i=1:1:length(POS1)
W=W+lambda(POS1(i))*Y(POS1(i))*P(:,POS1(i));
end
b=(W'*P(:,POS1(1))*Y(POS1(1))-1)*(-1)*Y(POS1(1));
%Projected training vectors
PSVMTRAINING=W'*P+b;
%The equation of the line on the plance X-Y, that seperates the
particular cluster with the rest is
%given as W(1)X+W(2)Y+b=0
%Projected testing vectors
figure
scatter(P(1,1:1:5),P(2,1:1:5),'s')
hold on
scatter(P(1,6:1:10),P(2,6:1:10),'d')
scatter(P(1,11:1:15),P(2,11:1:15),'v')
hold on
load TESTLPC
DATA=[];
for i=1:1:3
    DATA=[DATA TESTLPC{i}(1:1:x,:)];
end
U=isnan(DATA);
[R,S]=find(U==1);
for i=1:1:length(R)
    DATA(R(i),S(i))=eps;
end
P=E'*DATA;
PSVMTESTING=W'*P+b;
R=find(PSVMTESTING>0);
COL=[]
for i=1:1:length(R)
   F= find((CLUSTER-R(i))==0);
    COL=[COL F];
end
POS=(length(COL)/5)*100;
scatter(P(1,1:1:5),P(2,1:1:5),'*')
hold on
scatter(P(1,6:1:10),P(2,6:1:10),'o')
scatter(P(1,11:1:15),P(2,11:1:15),'+')
```

```
hold on
X=-3:0.1:3;
Y=(-b-W(1)*X)/W(2);
scatter(X,Y)
```

```
%svmforspeechkernel.m
%Dimensionality reduction using LDA before subjected to SVM
load TRAINLPC
s=[];
for i=1:1:3
    s=[s size(TRAINLPC{i},1)];
end
x=min(s);
DATA=[];
for i=1:1:3
    DATA=[DATA TRAINLPC{i}(1:1:x,:)];
end
U=isnan(DATA);
[R,S]=find(U==1);
for i=1:1:length(R)
    DATA(R(i),S(i))=eps;
end
SW=cov(DATA(:,1:1:5)',1)+cov(DATA(:,6:1:10)',1)+cov(DATA(:,11:1:15)',1);
[E1,V1]=eigs(SW,12);
for i=1:1:15
PR(:,i)=DATA(:,i)-E1*E1'*DATA(:,i);
end
M1=mean(PR(:,1:1:5)');
M2=mean(PR(:,6:1:10)');
M3=mean(PR(:,11:1:15)');
SB=cov([M1' M2' M3']',1);
[E,D]=eigs(SB,2);
%Projected vectors (P)
P=E'*DATA;
for i=1:1:size(P,2)
    for j=1:1:size(P,2)
        M(i,j)=k11(P(:,i),P(:,j),10);
    end
end
CLUSTER1=1:1:5;
Y=[ones(1,5) ones(1,10)*(-1)];
POS1=dosvmkernel(M,Y,P,x,E,CLUSTER1)
CLUSTER2=6:1:10;
Y=[ones(1,5)*(-1) ones(1,5) ones(1,5)*(-1)];
POS2=dosvmkernel(M,Y,P,x,E,CLUSTER2)
CLUSTER3=11:1:15;
```

```
Y=[ones(1,10)*(-1) ones(1,5)*(1)];
POS3=dosvmkernel(M,Y,P,x,E,CLUSTER3)

%dosvmkernel.m
function [POS]=dosvmkernel(M,Y,P,x,E,CLUSTER)
Y1=repmat(-1*Y,15,1);
M1=M.*Y1;
M2=[M1;Y];
N=[diag(M1)-1 ;0];
lambda=pinv(M2)*N;
[POS1,POS2]=find(lambda>=0);
S=0;
for i=1:1:size(P,2)
 S=S+lambda(i)*k11(P(:,i),P(:,POS1(1)),10)*Y(i)*Y(POS1(1));
end
b=Y(POS1(1))*(1-S);
%Projected training vectors
PSVMTRAINING=[];
for i=1:1:size(M,2)
PSVMTRAINING=[PSVMTRAINING sum(M(:,i).*lambda.*Y')+b];
end
load TESTLPC
DATA=[];
for i=1:1:3
    DATA=[DATA TESTLPC{i}(1:1:x,:)];
end
U=isnan(DATA);
[R,S]=find(U==1);
for i=1:1:length(R)
    DATA(R(i),S(i))=eps;
end
P1=E'*DATA;
for i=1:1:size(P1,2)
    for j=1:1:size(P,2)
P2(j,i)=k11(P1(:,i),P(:,j),10);
    end
end
PSVMTESTING=[];
for i=1:1:size(P2,2)
PSVMTESTING=[PSVMTESTING sum(P2(:,i).*lambda.*Y')+b];
end
R=find(PSVMTESTING>0);
COL=[]
for i=1:1:length(R)
   F= find((CLUSTER-R(i))==0);
    COL=[COL F];
```

```
end
POS=(length(COL)/5)*100;

%k11.m
function [res]=k11(x1,x2,c1)
x1=x1/sqrt(sum(x1.^2));
x2=x2/sqrt(sum(x2.^2));
c1=c1/2;
res=exp(-sum((x1-x2).^2)/(c1^2));
```

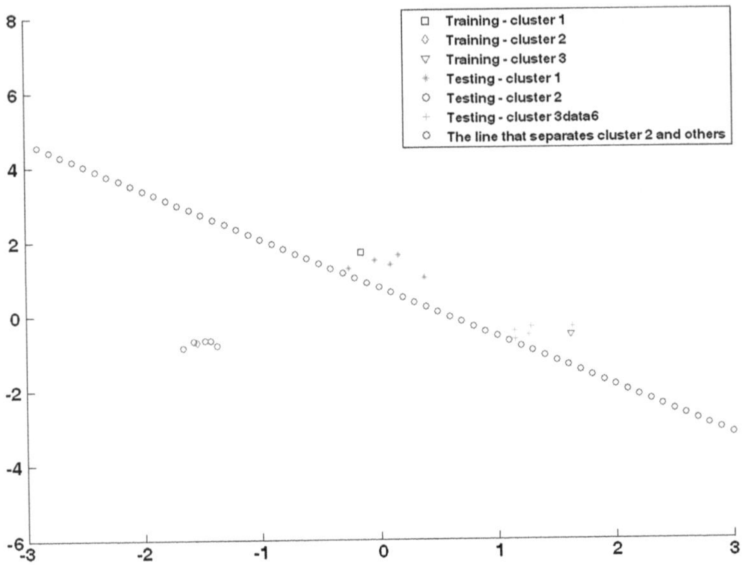

Fig. 1.9 Illustration of SVM—Part 2

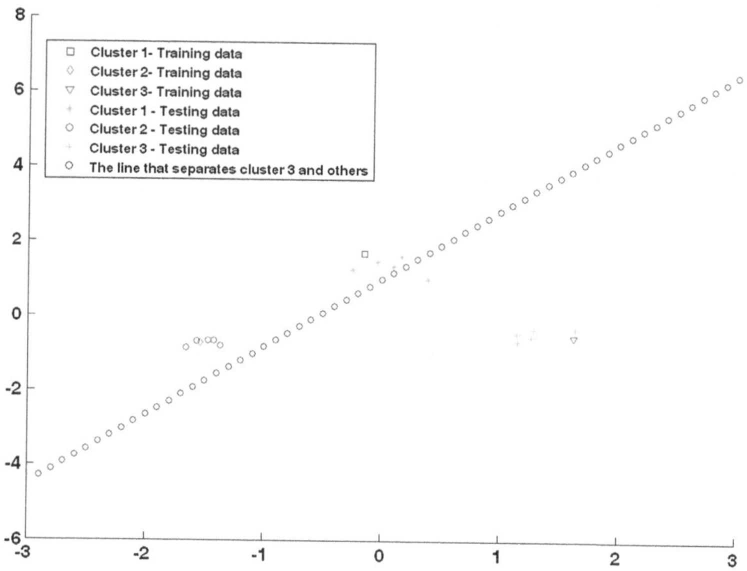

Fig. 1.10 Illustration of SVM—Part 3

1.4 Hidden Markov Model

The speech signal for the particular isolated word can be viewed as the one generated using the sequential generating probabilistic model known as hidden Markov model (HMM). Consider there are n states in the HMM. The particular isolated speech signal is divided into finite number of frames. Every frame of the speech signal is assumed to be generated from any one of the n states. Each state is modeled as the multivariate Gaussian density function with the specified mean vector and the covariance matrix. Let the speech segment for the particular isolated word is represented as vector S. The vector S is divided into finite number of frames (say M). The ith frame is represented as S_i. Every frame is generated by any of the n states with the specified probability computed using the corresponding multivariate Gaussian density model. The following assumptions are made in modeling the HMM (refer Fig. 1.11)

1. The first frame is assumed to be generated by the first state. The last frame is generated from the nth state.

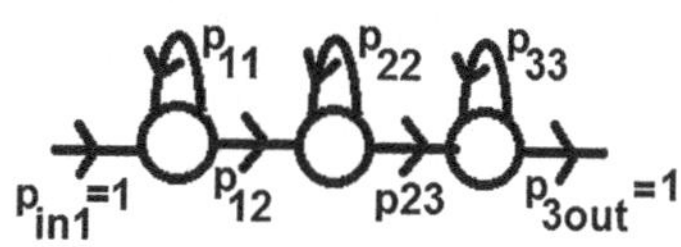

Fig. 1.11 Illustration of hidden Markov model with three states. Note that $p_{11} + p_{12} = p_{22} + p_{23} = p_{33} + p_{3\text{out}} = 1$

2. If the ith frame is generated from the state t, the $(i+1)$th frame is generated either from the state t or from the state $t+1$. This is known as first-order Markov model.

For the typical example of HMM with $n = 3$ and $M = 10$, the generating probability of the speech segment corresponding to the isolated word $S = [S_1\ S_2\ S_3\ S_4\ S_5\ S_6\ S_7\ S_8\ S_9\ S_{10}]$ is computed as follows. Let us assume that the frames S_1, S_2, S_3, and S_4 are generated from the state 1, the frames S_5, S_6, S_7, S_8 are generated from the state 2 and the frames S_9 and S_{10} are generated from the state 3. Thus, the probability of generating (generating probability) the isolated speech segment S is computed as follows:

1. The probability that the first frame is generated from the first state is 1 (assumption).
2. The probability of generating the arbitrary vector X generated from the ith state is given as $G_i(X)$.
3. The probability that ith frame is generated from the jth state is equal to the probability of transition of state associated with the $(i-1)$th frame to the state j associated with the ith frame. These probabilities are known as transition probabilities.
4. Let the transition probabilities are represented as p_{ij}. This is the probability that previous frame is generated from the ith state and the current frame is generated from the jth state.
5. Thus, the generating probability of the speech signal S is computed as $1 \times G_1(S_1)p_{11}G_1(S_2)p_{11}G_1(S_3)p_{11}G_1(S_4)p_{12}G_2(S_5)p_{22}G_2(S_6)p_{22}G_2(S_7)p_{22}$ $G_2(S_8)p_{23}G_3(S_9)p_{33}G_3(S_{10})$

HMM for the particular isolated word is completely described by state transition probabilities p_{ij} and conditional generating probability $G_i(X)$. The $G_i(X)$ is modeled as the Gaussian density function, which is described by its mean vector μ_i and covariance matrix C_i. The hidden parameters p_{ij}, μ_i, and C_i are obtained by maximizing the generating probability of that particular word as described below.

1. R number of speech segments corresponding to the particular isolated word are collected from the large number of speakers. R is the population size of the training set.
2. The length of the speech segments for the particular word is made identically equal using the preprocessing techniques (endpoint detection, dynamic time warping) described in this chapter. Let the kth speech segment (after preprocessing) corresponding to the particular isolated word be represented as WORD_k.
3. Every speech segment WORD_k is divided into M finite number of frames. Let WORD_{xy} be the yth frame from the xth speech segment. Let the size of the arbitrary frame WORD_{xy} be $f \times 1$.
4. Compute the squared distance between the consecutive frames for the xth speech segment as follows. Let the squared distance between the rth frame and the $(r+1)$th frame of the xth segment be computed as $d_{x,r,r+1} = (\text{WORD}_{x(r+1)} - \text{WORD}_{xr})^T(\text{WORD}_{x(r+1)} - \text{WORD}_{xr})$.
5. $D_x = [d_{x,1,2}\ d_{x,2,3}\ \ldots d_{x,(M-1),M}]^T$, for $x = 1, \ldots, R$.

6. Compute $D = \frac{1}{R}\sum_{i=1}^{i=R} D_i$.
7. Identify the number of elements in the vector D greater than the predefined threshold. This gives an estimate of the number of states of the HMM.
8. Thus, the number of states s is determined.
9. Let the forward probability of generating the particular vector sequence [WORD_{x1} $\text{WORD}_{x2} \ldots \text{WORD}_{xt}^{r}$] with the condition that tth frame is generated from the rth state. This probability is represented as $F(t, r)$. Note that x is the arbitrary speech segment.
10. Let the backward probability of generating the particular vector sequence [$\text{WORD}_{x(t+1)}$ $\text{WORD}_{x(t+2)} \ldots \text{WORD}_{xM}$] with the condition that tth frame is generated from the rth state. This probability is represented as $B(t, r)$. Note that x is the arbitrary speech segment.
11. The probability of generating the particular vector sequence [WORD_{x1} WORD_{x2} $\ldots \text{WORD}_{xt}^{r}$ $\text{WORD}_{x(t+1)} \ldots \text{WORD}_{xM}$] with the condition that tth frame is generated from the rth state is given as $F(t, r)B(t, r)$.
12. $F(1, 1)$ is the probability of generating the [WORD_{x1}^{1}] with the condition that first frame is generated from the first state. This is computed as $F(1, 1) = G_1(\text{WORD}_{x1})$.
13. $F_x(t, r)$ is the probability of generating the [WORD_{x1} $\text{WORD}_{x2} \ldots \text{WORD}_{xt}^{r}$] with the condition that tth frame is generated from the rth state. Also note that the $(t-1)$th can be either in the rth state or in the $(r-1)$th state. Hence, $F_x(t, r)$ is equivalently computed as the summation of

    - the probability of generating the [WORD_{x1} $\text{WORD}_{x2} \ldots \text{WORD}_{x(t-1)}^{r}$] with the condition that $(t-1)$th frame is generated from the rth state multiplied with the transition probability from rth state to rth state and the probability that the frame $\text{WORD}_{x(t)}$ is generated from the state r, given that tth frame is generated from state r;
    - the probability of generating the [WORD_{x1} $\text{WORD}_{x2} \ldots \text{WORD}_{x(t-1)}^{r-1}$] with the condition that $(t-1)$th frame is generated from the $(r-1)$th state multiplied with the transition probability from $(r-1)$th state to rth state and the probability that the frame $\text{WORD}_{x(t)}$ frame is generated from the state r, given that tth frame is generated from state r.

14. Hence, the recursive expression

$$F_x(t, r) = F_x(t-1, r-1)p_{r-1,r}G_r(\text{WORD}_{xt}) + F_x(t-1, r)p_{r,r}G_r(\text{WORD}_{xt}) \quad (1.34)$$

    is obtained.
15. $B_x(t, r)$ is the probability of generating the sequence [$\text{WORD}_{x(t+1)}$ $\text{WORD}_{x(t+2)}$ $\ldots \text{WORD}_{xM}$] with the condition that tth frame is generated from the rth state.
16. It is noted that the $(t+1)$th frame is generated either from the rth state or from the $(r+1)$th state. Hence, $B_x(t, r)$ can be equivalently computed as the summation of

- the product of the transition probability from rth state to rth state, the probability that the frame $\text{WORD}_{x(t+1)}$ is generated from the rth state, and the probability of generating the sequence [$\text{WORD}_{x(t+2)}$ $\text{WORD}_{x(t+3)} \ldots$ WORD_{xM}] with the condition that $(t+1)$th frame is generated from the rth state;
- the product of the transition probability from rth state to $(r+1)$th state, the probability that the frame $\text{WORD}_{x(t+1)}$ is generated from the $(r+1)$th state, and the probability of generating the sequence [$\text{WORD}_{x(t+2)}$ $\text{WORD}_{x(t+3)} \ldots$ WORD_{xM}] with the condition that $(t+1)$th frame is generated from the $(r+1)$th state.

17. Hence, the recursive equation

$$B_x(t,r) = p_{r,r}G_r(\text{WORD}_{x(t+1)})B_x(t+1,r) + p_{r,r+1}G_{r+1}(\text{WORD}_{x(t+1)})B(t+1,r+1) \tag{1.35}$$

is obtained.

18. Also note that $B_x(M,s)$ is the probability of generating [WORD_{x1} $\text{WORD}_{x2} \ldots$ WORD^s_{xM}] with the condition that Mth frame is generated from the sth state. As the probability of generating the Mth frame from the state s is 1, $B_x(M,s)$ is the probability of generating the sequence [WORD_{x1} $\text{WORD}_{x2} \ldots \text{WORD}^s_{xM}$].
19. Using $F_x(t,r)$ and $B_x(t,r)$, the unknown parameters describing the HMM for the particular isolated word segment are determined using Baum–Welch technique as described below.

1.4.1 Baum–Welch Technique to Obtain the Unknown Parameters in HMM

The Baum–Welch technique to obtain the unknown parameters in HMM is described as follows. It is assumed that any arbitrary frame is generated from any one among the s states. We have $M \times R$ frames available from the training set to formulate the unknown parameters of the HMM. From the above discussion, we understand that the probability of generating the particular vector sequence [WORD_{x1} $\text{WORD}_{x2} \ldots \text{WORD}^r_{xt}$ $\text{WORD}_{x(t+1)} \ldots \text{WORD}_{xM}$] with the condition that tth frame is generated from the rth state is given as $F_x(t,r)B_x(t,r)$. The tth frame can be generated only from any of the s states. Hence, the fraction of the tth frame of the xth speech segment, being generated from the rth state, is computed as

$$A_x(t,r) = \frac{F_x(t,r)B_x(t,r)}{\sum_{k=1}^{k=s} F_x(t,k)B_x(t,k)}. \tag{1.36}$$

It is noted that summation of the (1.36) over the range $r = 1, \ldots, s$ is equal to 1. Hence, it is the fraction of the tth frame being generated from the rth state. The total number of frames (out of $M \times R$) generated from the rth state is computed as follows:

$$\mathrm{NF}_r = \sum_{x=1}^{x=R}\sum_{t=1}^{t=M} A_x(t,r) \tag{1.37}$$

It is noted that $\sum_{r=1}^{r=s} \mathrm{NF}_r = M \times R$. The mean vector and the covariance matrix of the kth state are computed as follows:

1. Every vector WORD_{xt} is assumed to be generated from any of the s states. Let the probability that the frame WORD_{xt} is generated from the state k be represented as g_{xtk}.
2. Using g_{xtk}, the mean vector of the kth state is computed as

$$\mathrm{mean}_k = \sum_{x=1}^{x=R}\sum_{t=1}^{t=M} \mathrm{WORD}_{xt}\ g_{xtk} \tag{1.38}$$

3. Similarly, the covariance matrix of the kth state is computed as

$$\mathrm{covar}_k = \sum_{x=1}^{x=R}\sum_{t=1}^{t=M} [\mathrm{WORD}_{xt} - \mathrm{mean}_k][\mathrm{WORD}_{xt} - \mathrm{mean}_k]^T g_{xtk} \tag{1.39}$$

4. The probability g_{xtk} is computed as $\frac{\text{Fractional contribution of the frame } \mathrm{WORD}_{xt} \text{ generated from the } k\text{th state}}{\text{Total number of frames generated from the } k\text{th state}}$. Thus, g_{xtk} is computed as follows:

$$g_{xtk} = \frac{A_x(t,k)}{\sum_{x=1}^{x=R}\sum_{t=1}^{t=M} A_x(t,k)} \tag{1.40}$$

5. The transition probability p_{ij} is the probability that the frame is changing from the ith state to the jth state. This is computed as the ratio of the number of frames generated by the state i, which is followed by the frame generated from the jth state, to the number of frames generated from the state i.
6. The probability that the tth frame is generated from the ith state and $(t+1)$th frame is generated from the $(i+1)$th state in the xth speech segment is computed as the probability of generating the xth speech segment with tth frame generated from the ith state and $(t+1)$th frame generated from the $(i+1)$th state and is represented as $F_x(t,i)p_{i(i+1)}G_j(\mathrm{WORD}_{x(t+1)})F_x(t+1,i+1)$. As discussed earlier, we assume that every frame of the xth speech segment contributes to the event that it is generated from the ith state and the next frame is generated from the $(i+1)$th state. Thus, for the xth speech segment, the fraction of the tth frame generated from the ith state and the $(t+1)$th frame generated from the $(i+1)$th state is given as

$$B_x(t,i,j) = \frac{F_x(t,i)p_{ij}G_j(\mathrm{WORD}_{x(t+1)})F_x(t+1,j)}{\sum_{k=1}^{s-1}\sum_{l=k}^{k+1} F_x(t,k)p_{kl}G_j(\mathrm{WORD}_{x(t+1)})F_x(t+1,l)} \quad (1.41)$$

7. It is noted that $\sum_{k=1}^{i=s-1}\sum_{l=k}^{i=k+1} B_x(t,k,l) = 1$. Also note that in our model, j can take either i or $i+1$ for particular i in (1.41).
8. Thus, the number of frames generated by the state i, followed by the frame generated from the jth state, is computed as

$$\sum_{x=1}^{x=R}\sum_{t=1}^{t=M} B_x(t,i,j) \quad (1.42)$$

9. Thus, the probability that the frame is changing from the ith state to the jth state (transition probability) is computed as

$$p_{ij} = \frac{\sum_{x=1}^{x=R}\sum_{t=1}^{t=M} B_x(t,i,j)}{\sum_{x=1}^{x=R}\sum_{t=1}^{t=M} A_x(t,i)} \quad (1.43)$$

10. It is also noted from the Eqs. (1.41) and (1.43) that the transition probability p_{ij} is present in both sides of the equation. Hence, obtaining the unknown parameters mean_k, covar_k, and the transition probability p_{ij} using the above set of equations is done using the method "Expectation–Maximization" (refer Appendix B) algorithm as described below.

1.4.2 Steps to Compute the Unknown Parameters of HMM Using Expectation–Maximization Algorithm

1. Initialize the number of states s, mean vectors mean_k, and covariance matrices covar_k for $k = 1, \ldots, s$ and the valid transition probabilities ($p_{ii} + p_{i(i+1)} = 1$ and $p_{ij} = 0$ with $j \neq i$ or $j \neq (i+1)$).
2. Compute $F_x(1,1)$ using $G_1(\mathrm{WORD}_{x1})$ and $B_x(M,s)$ using the Viterbi algorithm described in Sect. 1.4.1.
3. Compute $F_x(t,r)$ and $B_x(t,r)$ for $t = 1, \ldots, M$; $r = 1, \ldots, s$ using (1.34) and (1.35).
4. Compute the updated mean vectors mean_k and covariance matrices covar_k for $k = 1, \ldots, s$ and the transition probability p_{ij} using (1.38), (1.39), and (1.40), respectively.
5. Compute the generating probability of all the training sequences of the current iteration, which is computed as $gp(\mathrm{iter}) = \sum_{x=1}^{x=R} B_x(M,s)$. The generating probability.

6. Repeat steps 2 and 3 until $(gp(\text{iter}+1) - gp(\text{iter}))^2$ is less than the predefined threshold value T_h.

1.4.3 Viterbi Algorithm to Compute the Generating Probability of the Arbitrary Speech Segment $WORD_X$

1. Formulate the matrix grid MG (refer Fig. 1.12) with the number of rows equal to the number of states s and the number of columns equal to the number of frames (M) of the speech segment WORD_X.
2. We need to find the path from the point $(1, 1)$ to (s, M) that gives the best alignment based on the calculation of maximum generating probability.
3. The first frame is assumed to be generated from the state 1. Hence, the generating probability of the WORD_{X1} is computed as $G_1(\text{WORD}_{X1})$.
4. MG(1, 1) is filled up with the value $G_1(\text{WORD}_{X1})$. $\text{MG}(i, 1) = 0\,\forall i = 2, \ldots, s$. The second frame is generated either from the state 1 or from the state 2 (from basic assumptions). If the second frame is generated from the state 1, the generating probability of the sequence up to the second frame is computed as $\text{MG}(1, 1)p_{11}G_1(\text{WORD}_{X1})$. Thus, $M(1, 2) = \text{MG}(1, 1)p_{11}G_1\text{WORD}_{X2}$. Similarly, $M(2, 2) = \text{MG}(1, 1)p_{12}G_2(\text{WORD}_{X2})$. $\text{MG}(i, 2) = 0\ \forall\ i = 3, \ldots, s$.
5. MG(1, 3) is the best path up to frame 3 generated from the state 1. This is computed as $\text{MG}(1, 3) = \text{MG}(1, 2)p_{11}G_1(\text{WORD}_{X3})$
6. MG(2, 3) is the best path up to frame 3 generated from the state 2. This path can come either from MG(1, 2) or from MG(2, 2). Hence, the best among them is chosen as the best generating probability value for MG(2, 3), which is computed as $\max(\text{MG}(1, 2)p_{12}G_2(\text{WORD}_{X3}),\ \text{MG}(2, 2)p_{22}G_2(\text{WORD}_{X3}))$.
7. Similarly, $\text{MG}(3, 3) = \text{MG}(2, 2)p_{23}G_3(\text{WORD}_{X3})$. Note that $\text{MG}(i, 3) = 0\,\forall\, i = 4, \ldots, s$.
8. The generalized expression for computing the content of the matrix MG is given below.

$$\text{MG}(t, i) = \max(\text{MG}(t-1, i-1)p_{(t-1)t}G_t(\text{WORD}_{Xi}), \text{MG}(t, i-1)p_{tt} \\ G_t(\text{WORD}_{Xi})), \quad \forall\, i < t$$
$$\text{MG}(t, t) = \text{MG}(t-1, t-1)p_{(t-1)t}G_t(\text{WORD}_{Xt}), \quad \forall\, t \le s$$
$$\text{MG}(t, i) = 0 \quad \forall\, t > i$$
$$\text{MG}(t, i) = \text{MG}(t, i-1)p_{tt}G_t(\text{WORD}_{Xi}), \quad \forall\, i > s$$

9. Once MG matrix is completely filled, backtrace the best path from $\text{MG}(s, M)$ to MG(1, 1) so that the generating probability is maximized. This helps to obtain the alignment of individual frames with the corresponding states (used in Sect. 1.4.5). $\text{MG}(s, M)$ gives directly the generating probability of the corresponding isolated word.

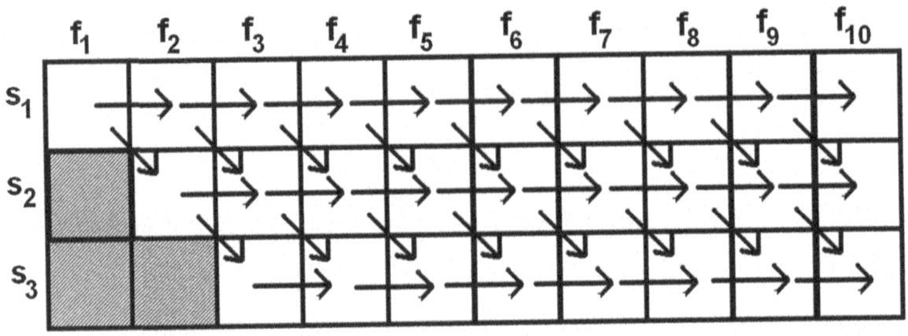

Fig. 1.12 Sample matrix grid (MG) of arbitrary isolated word to compute the generating probability using Viterbi algorithm. The *arrows* in the figure show the possible transition of states between frames. MG(s, M) gives the generating probability of the arbitrary isolated word

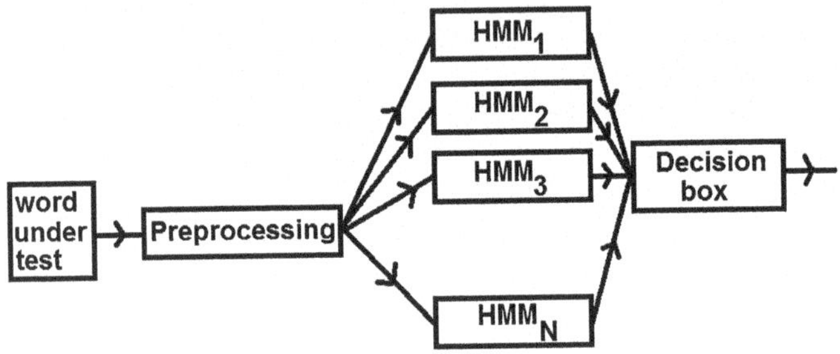

Fig. 1.13 Illustration of isolated word recognition using HMM. HMM$_r$ is the hidden Markov model for the rth isolated word

1.4.4 Isolated Word Recognition Using HMM

HMMs are obtained for the finite number of isolated words (dictionary words) that are to be classified. The number of HMMs is equal to the number of isolated words. Given the unknown speech signal (test signal), the following steps are followed to classify it as one among the dictionary words.

1. Preprocess the test signal $\text{WORD}_{\text{test}}$ using the endpoint detection and the dynamic time warping to make the number of frames exactly equal to the number of frames of the individual isolated training words.
2. The generating probabilities of the preprocessed test signal ($B_{\text{test}}(M, s)$) using all the HMMs stored in the database are computed. The test signal is declared as the word corresponding to the HMM (refer Fig. 1.13) that gives the maximum generating probability.
3. Note that the generating probability is computed using the Viterbi algorithm as described in Sect. 1.4.3.

1.4.5 Alignment Method to Model HMM

The alternate simple method to obtain the unknown parameters mean_k, covar_k, *and* p_{ij} are obtained using Viterbi technique as follows:

1. Let the training isolated speech segments (for the identical words) after pre-processing (endpoint detection and DTW) be represented as WORD_1, WORD_2, $\ldots$, WORD_R.
2. Initialize the number of states s and the valid mean_k, covar_k, *and* p_{ij} of the HMM, iter $= 0$ and $gp(1) = 100$.
3. Let iter $=$ iter $+ 1$. For every training set, obtain the best alignment using Viterbi algorithm as described in Sect. 1.4.3 using the latest HMM.
4. Thus, the individual frames of all the training data are assigned to the corresponding states.
5. Compute empirically the mean of all the frames assigned to the individual state (say k) to obtain mean_k and covar_k.
6. Compute the number of frame transition from the state i to state j. Let it be n_{ij}. Also calculate the number of frames generated from the state i. Let it be n_i. Thus, the transition probability p_{ij} is calculated empirically as $p_{ij} = \frac{n_{ij}}{n_j}$. Thus, the parameters are updated to obtain the latest HMM.
7. Compute the summation of the generating probability of the individual training data (refer Sect. 1.4.3) using the updated HMM. Let it be $gp(\text{iter} + 1)$.
8. Repeat the steps 2–5 until $(gp(\text{iter} + 1) - gp(\text{iter}))^2$ is less than the predefined threshold t_h.
9. Thus, the final parameters mean_k, covar_k, *and*p_{ij} describe the HMM for the corresponding isolated word.

1.4.6 Illustration of Hidden Markov Model

The LPC data (182 coefficients) are collected from three words. The 10 isolated speech segments correspond to every words. 50 % of the collected data are used to model HMM using Baum–Welch method. The remaining fifty percent data are used for testing. The percentage of correct classification is achieved as 86.67 % (13 out of 15). As the computation involves very less values, the computation is performed using logarithmic scale. In particular, for the arbitrary probabilities $p1$ and $p2$, $\log(p1 + p2)$ is computed as follows. If $p1 > p2$, compute $x = \log(p1) - \log(p2)$. Compute $\log(p1 + p2)$ as $\log(p1) + \log(1 + \frac{p2}{p1})$, in which $\log(1 + \frac{p2}{p1})$ is computed as $\log(1 + e^{-x})$ and is approximately computed as $\frac{\log(2)}{(1+0.4x+0.97(x^2))}$.

```
%hmmforspeech.m
%TRAINLPC{1} is the LPC speech data collected from the word 1. Each column
 of TRAINLPC{1} belongs to the individual isolated speech.
%Formulating Hidden Markov Model
clear all
close all
load TRAINLPC
s=[];
for i=1:1:3
    s=[s size(TRAINLPC{i},1)];
end
mx=min(s);
DATA=[];
for i=1:1:3
    DATA=[DATA TRAINLPC{i}(1:1:mx,:)];
end
DATA1=[];
for i=1:1:size(DATA,2)
    temp=DATA(:,i);
DATA1=[DATA1 reshape(temp,13,14)];
end
[E,D]=eigs(cov(DATA1'),6);
DATA1=E'*DATA1;
%Initializing the mean and the co-variance matrix for 5 states.
S=5;
ITER=50;
[M,COV,INDEX,COL]=kmeansforspeech(DATA1,S,ITER);
%Initializing the transition probabilities (p) for 5 states
S=length(COV);
for i=1:1:S-1
r=rand;
p(i,:)=[zeros(1,i-1) r 1-r  zeros(1,S-1-i)];
end
p=[p;zeros(1,size(p,2)-1) 1];
S=length(M);
%HMM for the word 1
%F(m,n) is the probability of generating the word 'WORD1' upto 'm' frame
%with mth frame is generated from the nth state.
F=zeros(14,S);
R=zeros(13,S);
R=[R; ones(1,S)];
for word=1:1:3
DATA=DATA1(:,(word-1)*70+1:1:(word-1)*70+70);
for iteration=1:1:100
    iteration
for w=1:1:5
WORD1=DATA(:,(w-1)*14+1:1:(w-1)*14+14);
F(1,1)=abs(gaussvalue(WORD1(:,1),M{1},COV{1}));
for i=2:1:13
F(i,1)=logsum(F(i-1,1),p(1,1)*gaussvalue(WORD1(:,i),M{1},COV{1}));
for j=2:1:S
F(i,j)=logsum(logsum(F(i-1,j-1),p(j-1,j)*gaussvalue(WORD1(:,i),M{j},...
COV{j})),F(i-1,j)+p(j,j)*gaussvalue(WORD1(:,3),M{j},COV{j}));
end
end
F(14,S)=logsum(F(13,S-1),p(S-1,S)*gaussvalue(WORD1(:,14),M{S},COV{S}));
FF{w}=F;
%R(m,n) is the probability of generating the word WORD1 from (m+1)th frame
%to the final frame with the condition mth frame is generated from the nth
```

```
%state.
R(13,S)=abs((p(S,S)*gaussvalue(WORD1(:,14),M{S},COV{S})));
for i=12:-1:1
R(i,S)=logsum(log(p(S,S)*gaussvalue(WORD1(:,i+1),M{S},COV{S})),R(i+1,S));
            for j=S-1:-1:1
R(i,j)=logsum(logsum(p(j,j)*gaussvalue(WORD1(:,i+1),M{j},COV{j}),...
R(i+1,j)),logsum(p(j,j+1)*...
gaussvalue(WORD1(:,i+1),M{j+1},COV{j+1}),R(i+1,j+1)));
end
end
RR{w}=R;
for i=1:1:13
    for j=1:1:S
        num=logsum(F(i,j),R(i,j));
        den=0;
        for k=1:1:S
            den=logsum(den,logsum(F(i,k),R(i,k)));
        end
        A(i,j)=exp(num-den);
    end
end
for i=1:1:size(A,1)
A(i,:)= A(i,:)/sum(A(i,:));
end
AA{w}=A;
for i=1:1:13
    for j=1:1:S-1
        for k=j:1:j+1
            num=logsum(logsum(F(i,j),p(j,k)*gaussvalue(WORD1(:,i+1),...
            M{j+1},COV{j+1})),F(i+1,k));
            den=0;
            for l=1:1:S-1
                for m=l:1:l+1
            den=logsum(den, logsum(logsum(F(i,l),p(l,m)...
              *gaussvalue(WORD1(:,i+1),M{l+1},COV{l+1})),F(i+1,m)));
                  end
              end
         B(i,j,k)=exp(num-den);
    end
    end
for i=1:1:size(B,1)
    s=0;
for j=1:1:size(B,2)
    for k=1:1:size(B,3)
s=s+B(i,j,k);
    end
end
B(i,:,:)=B(i,:,:)/s;
end
BB{w}=B;
end
end

for i=1:1:S-1
    for j=i:1:i+1
        num=0;
        den=0;
for k=1:1:5
    for l=1:1:13
```

```
                num=logsum(num,BB{k}(1,i,j));
                den=logsum(den,AA{k}(1,i));
            end
        end
        p(i,j)=exp(num-den);
    end
end
for i=1:1:size(p,1)
p(i,:)= p(i,:)/sum(p(i,:));
end

for k=1:1:S
DEN=0;
for t=1:1:13
for x=1:1:5
DEN=DEN+AA{x}(t,k);
end
end
for x=1:1:5
    for t=1:1:13
m(x,k,t)=AA{x}(t,k)/(DEN+eps);
end
end
end
for k=1:1:S
MEAN{k}=0;
COV{k}=0;
for x=1:1:5
    WORD=DATA(:,(x-1)*14+1:1:(x-1)*14+14);
    for t=1:1:13
    MEAN{k}=MEAN{k}+m(x,k,t)*WORD(:,t);
    end
end
for x=1:1:5
    WORD=DATA(:,(x-1)*14+1:1:(x-1)*14+14);
    for t=1:1:13
    COV{k}=COV{k}+m(x,k,t)*(WORD(:,t)-MEAN{k})...
    *(WORD(:,t)-MEAN{k})';
    end
end
end

end
HMMMODEL{word}{1}=p;
HMMMODEL{word}{2}=COV;
HMMMODEL{word}{3}=M;
end
save HMMMODEL mx E

%hmmtestingforspeech
clear all
close all
load HMM
load TESTLPC
DATA=[];
for i=1:1:3
    DATA=[DATA TESTLPC{i}(1:1:mx,:)];
end
DATA1=[];
for i=1:1:size(DATA,2)
    temp=DATA(:,i);
DATA1=[DATA1 reshape(temp,13,14)];
end
```

```
DATA1=E'*DATA1;
%Generating probability of the first word
S=5;
F=14;
G=[];
INDEX=[];
for x=1:1:15
DATA=DATA1(:,(x-1)*F+1:1:(x-1)*F+F);
GP=[];
for w=1:1:3
p= HMMMODEL{w}{1};
M=zeros(5,14);
M(1,1)=gaussvalue(DATA(:,1),HMMMODEL{w}{3}{1},...
HMMMODEL{w}{2}{1});
%Alignment using viterbi algorithm
for j=2:1:F
M(1,j)=M(1,j-1)*p(1,1)*gaussvalue(DATA(:,j),...
HMMMODEL{w}{3}{1},HMMMODEL{w}{2}{1});
end
%M(2,2)=M(1,1)*p(1,2)*gaussprob(DATA(:,2),...
HMMMODEL{w}{2}{2},HMMMODEL{w}{3}{2});
for i=2:1:S
for j=2:1:i
M(j,i)=max( M(j-1,i-1)*p(j-1,j)*gaussvalue(DATA(:,i),...
HMMMODEL{w}{3}{j},HMMMODEL{w}{2}{j}),...
M(j,i-1)*p(j,j)*gaussvalue(DATA(:,i),HMMMODEL{w}{3}{j},...
HMMMODEL{w}{2}{j}));
end
end
for i=S+1:1:F
    for j=2:1:S
M(j,i)=max( M(j-1,i-1)*p(j-1,j)*gaussvalue(DATA(:,i),...
HMMMODEL{w}{3}{j},HMMMODEL{w}{2}{j}),...
M(j,i-1)*p(j,j)*gaussvalue(DATA(:,i),HMMMODEL{w}{3}{j},...
HMMMODEL{w}{2}{j}));
end
end
%Backtrace to compute generating probability
trace=[5];
Q=5;
for i=(F-1):-1:1
if(Q==1)
trace=[trace 1];
else
[Y,Z]=max([M(Q,i) M(Q-1,i)]);
trace=[trace Q-Z+1];
Q=Q-Z+1;
end
end
o=trace(length(trace):-1:1);
gp=gaussvalue(DATA(:,1),HMMMODEL{w}{3}{o(1)},HMMMODEL{w}{2}{o(1)});
for i=2:1:F
gp=gp*p(o(i-1),o(i))*gaussvalue(DATA(:,i),HMMMODEL{w}{3}{o(i)},...
HMMMODEL{w}{2}{o(i)});
end
GP=[GP gp];
    end
[V,I]=max(GP);
G=[G;GP];
INDEX=[INDEX I];
end
%kmeansforspeech.m
function [M,COV,INDEX,COL,P1]=kmeansforspeech(DATA,n,N)
s1=size(DATA,1);
s2=size(DATA,2)/n;
```

```
r=0;
M=[];
for i=1:1:n
M{i}=mean(DATA(:,(i-1)*s2+1:1:(i-1)*s2+s2)')';
end
COL=[];
for iteration=1:1:N
clear M1
for k=1:1:n
    C{k}=[];
    INDEX{k}=[];
end
count=0;
k=1;
M1=[];
for i=1:1:n
    if(length(M{i})==size(DATA,1))
count=count+1;
M1{k}=M{i};
k=k+1;
    end
end
n=count;
clear M
M=M1;
S=0;
for i=1:1:size(DATA,2)
    D=[];
   for j=1:1:n
    D=[D sum((DATA(:,i)-M{j}).^2)];
   end
[P,Q]=min(D);
C{Q}=[C{Q} DATA(:,i)];
INDEX{Q}=[INDEX{Q} i];
S=S+P;
end
COL=[COL S];
for i=1:1:n
M{i}=mean(C{i}')';
end
end
for i=1:1:n
    COV{i}=cov(C{i}');
end
for i=1:1:n
P1(i)=length(INDEX{i})/size(DATA,2);
end
%logsum.m
%fuction that computes log(p1+p2)
function [res]=logsum(p1,p2)
p1=abs(p1)+eps;
p2=abs(p2)+eps;
%logsum(p1+p2) is computed
if (p1>p2)
    x=log(p1)-log(p2);
    res=log(p1)+log(2)/(1+0.4*x+0.97*(x^2));
else
    x=log(p2)-log(p1);
    res=log(p2)+log(2)/(1+0.4*x+0.97*(x^2));
end
res=abs(res);
\begin{verbatim}
```

```
function [res]=gaussvalue(x,m,c)
t=(x-m)'*pinv(c)*(x-m);
t1=exp(-t/2);
t2=1/((2*pi)^(length(x)/2));
[E,V]=eig(c);
V=diag(V);
[P,Q]=find(V>0);
t3=1;
for i=1:1:length(P)
    t3=t3*V(P(i));
end
t3=1/(abs(t3)^(1/2));
res=t1*t2*t3;
```

1.5 Gaussian Mixture Model

The HMM is the sequential generating probabilistic model. The model assumes that the correlation between the adjacent frames is high. The model works better if the adjacent frames are highly correlated. This is true for the isolated speech frames. But the algorithm to obtain the model is complex. The Gaussian mixture model (GMM) is the alternate generating probabilistic model, which assumes that all the frames of the particular word are generated from the linear combinations of multivariate Gaussian density model. Let $P(\mathrm{WORD}_{xt})$ be the tth frame of the isolated word WORD_x. The probability of generating the frame $P(\mathrm{WORD}_{xt})$ using GMM is computed as follows:

$$P_{\mathrm{GMM}}(\mathrm{WORD}_{xt}) = \sum_{k=1}^{k=s} c_k G_k(\mathrm{WORD}_{xt}) \tag{1.44}$$

where s is the number of mixtures, c_k is the probability of the kth mixture, and G_k is the multivariate Gaussian density function with mean vector mean_k and covariance matrix covar_k. It is noted that $G_k(\mathrm{WORD}_{xt})$ is the probability of generating the frame WORD_{xt} from the kth mixture. The unknown parameters are c_k, mean_k, covar_k, and s. The unknown parameters are obtained such that the generating probability of the isolated word using GMM is maximized as described below. Note that the $\mathrm{WORD}_x \ \forall \ x = 1, \ldots, R$ belongs to the identical isolated words.

1. Initialize the number of mixtures s. Initialize the mean vectors mean_k and the covariance matrices covar_k. These are obtained using the unsupervised clustering technique like k-means algorithm (refer Sect. 1.6.2). k-means algorithm helps in identifying the individual frames belonging to the corresponding mixtures. The mean vectors are empirically calculated using the frames assigned to the corresponding mixtures using the k-means algorithm. Similarly, covariance matrices are empirically calculated using the frames assigned to the corresponding mixtures using the k-means algorithm. The probability of the kth mixture c_k's is obtained

$$\frac{\text{Total number of frames assigned to the mixture } k}{\text{Total number of frames in the training set for the isolated word } \mathrm{WORD}_x = RM}.$$

2. All the frames are assumed to be generated from all the mixtures. Hence, the best optimal values of the unknown parameters are computed by maximizing the generating probability of that particular word as follows.
3. Fractional number of the frame WORD_{xt} generated from the mixture r is given as follows:

$$A_x(t, r) = \frac{c_r G_r(\mathrm{WORD}_{xt})}{\sum_{k=1}^{k=s} c_r G_r(\mathrm{WORD}_{xt})} \tag{1.45}$$

4. It is noted that $\sum_{r=1}^{r=s} A(t, r) = 1$.
5. The mean of the kth mixture is given as the weighted mean of all the frames used to model GMM.

$$\mathrm{mean}_k = \sum_{x=1}^{x=R} \sum_{t=1}^{t=M} w_{xtk} \mathrm{WORD}_{xt} \tag{1.46}$$

6. The weight w_{xtk} is the probability of the frame WORD_{xt} generated from the kth mixture. This is computed as the ratio $\frac{\text{Fractional contribution of the frame } \mathrm{WORD}_{xt} \text{ generated from the } k\text{th mixture}}{\text{Total number of frames generated from the } k\text{th mixture}}$.
7.

$$w_{xkt} = \frac{A_x(t, k)}{\sum_{x=1}^{x=R} \sum_{t=1}^{t=M} A_x(t, r)} \tag{1.47}$$

8. Similarly, the covariance matrix of the kth mixture is given as the weighted mean of all the frames used to model GMM.

$$\mathrm{covar}_k = \sum_{x=1}^{x=R} \sum_{t=1}^{t=M} w_{xtk} [\mathrm{WORD}_{xt} - \mathrm{mean}_k]^T [\mathrm{WORD}_{xt} - \mathrm{mean}_k] \tag{1.48}$$

9. c_r is the probability of the state r, which is computed as the ratio $\frac{\text{Number of frames generated from the } k\text{th mixture}}{\text{Total number of frames}}$

$$c_r = \frac{\sum_{x=1}^{x=R} \sum_{t=1}^{t=M} A_x(t, r)}{\sum_{x=1}^{x=R} \sum_{t=1}^{t=M} \sum_{k=1}^{k=s} A_x(t, k)} = \frac{N_r}{N} \tag{1.49}$$

where N_r is the number of frames generated from the state r and N is the total number of frames used to model GMM $=$ MR.

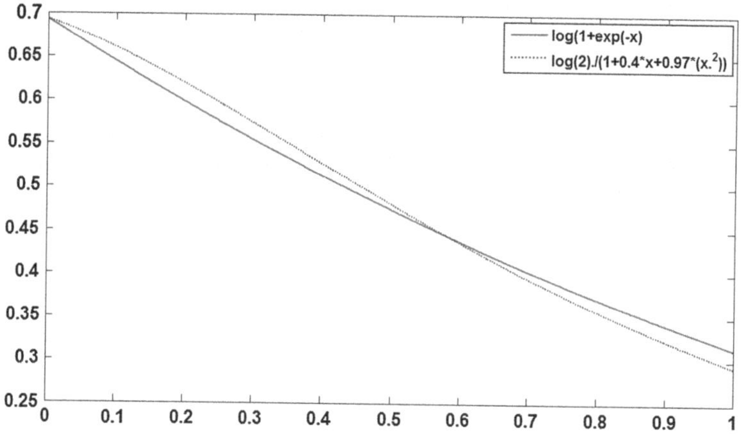

Fig. 1.14 Illustration of the approximation of the function $\log(1 + \exp(-x))$ over the period 0 to 1

1.5.1 Steps Involved to Model GMM Using Expectation–Maximization Algorithm

1. Initialize the parameters mean_k, covar_k, and c_k, the iteration number iter $= 0$ and $gp(1) = 100$.
2. Compute iter $=$ iter $+ 1$.
3. Compute $A_x(t, r)$ using (1.45).
4. Compute w_{xkt} using (1.47).
5. Compute the unknown parameters mean_k, covar_k, and c_k using (1.46), (1.48), and (1.49), respectively.
6. Compute the generating probability of the word WORD_x as $\prod_{t=1}^{t=M} P_{\text{GMM}}(\text{WORD}_{xt})$. Thus, summation of the generating probability of all the training sets is computed as $\sum_{x=1}^{x=M} \prod_{t=1}^{t=M} P_{\text{GMM}}(\text{WORD}_{xt})$. Let it be the generating probability of the training set for the current iteration represented as $gp(\text{iter} + 1)$. The values obtained using the above computation are normalized such that summation of the generating probability of the particular word from all the model is 1.
7. Repeat steps (2)–(6) until $(gp(\text{iter} + 1) - gp(\text{iter}))^2$ is less than the predefined threshold value t_h.
8. Note that the step (4) is the expectation stage and the step (5) is the maximization stage of the Expectation–Maximization algorithm (refer Appendix B).

1.5.2 Isolated Word Recognition Using GMM

GMMs are obtained for the finite number of isolated words (dictionary words) that are to be classified. The number of GMMs is equal to the number of isolated words.

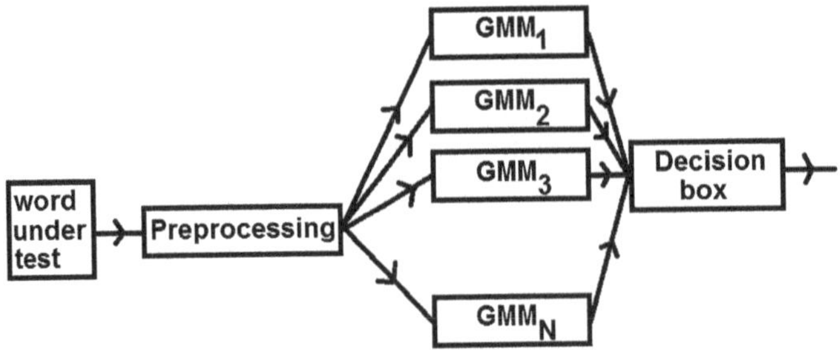

Fig. 1.15 Illustration of isolated word recognition using GMM. GMM_r is the Gaussian mixture model for the rth isolated word

Given the unknown speech signal (test signal), the following steps are followed to classify it as one among the dictionary words.

1. Preprocess the test signal $WORD_{test}$ using the endpoint detection and the dynamic time warping to make the number of frames exactly equal to the number of frames of the individual isolated training words.
2. The generating probabilities of the preprocessed test signal using all the GMMs stored in the database are computed. The test signal is declared as the word corresponding to the GMM (refer Fig. 1.15) that gives the maximum generating probability.

1.5.3 Illustration of GMM

HMM-based speech recognition is illustrated with three words. Ten data for every isolated word (i.e., 30 data) are collected. 50 % of the collected data are used for training, and the remaining 50 % data are used for testing. Every isolated word is divided into frames of 30 ms, and 13 LPC are collected for every frame. There are totally 14 frames for every speech segment. Principal component analysis (PCA) is used for dimensionality reduction to obtain six coefficients for every frame (refer Sect. 1.7.1). Using the obtained lower-dimensional vectors, GMM is obtained for every word as GMM1, GMM2, and GMM3 using the method described in Sect. 1.5.2. The number of states in GMM is chosen as 5. The initial values for the mean, probabilities and covariance matrices of the individual states are obtained using k-means algorithm (refer Sect. 1.6.2). The generating probability is calculated for the first testing data using the three models GMM1, GMM2, and GMM3. Identify the model that has maximum generating probability (say n). Declare the first data belong to the nth word. This is repeated for the remaining words. The typical generating probability obtained for the 15 testing words is tabulated below for better illustration. The percentage of success in correct classification is obtained as 100 % (15 out of 15).

```
%gmmforspeech.m
clear all
close all
load TRAINLPC
s=[];
for i=1:1:3
    s=[s size(TRAINLPC{i},1)];
end
mx=min(s);
DATA=[];
for i=1:1:3
    DATA=[DATA TRAINLPC{i}(1:1:mx,:)];
end
DATA1=[];
for i=1:1:size(DATA,2)
    temp=DATA(:,i);
DATA1=[DATA1 reshape(temp,13,14)];
end
[E,D]=eigs(cov(DATA1'),6);
DATA1=E'*DATA1;
%Initializing the mean and the co-variance matrix for 5 states.
S=5;
ITER=100;
F=14;
%MODEL THE FIRST WORD
for word=1:1:3
DATA=DATA1(:,(word-1)*70+1:1:(word-1)*70+70);
[M,COV,INDEX,COL,P]=kmeansforspeech(DATA,S,ITER);
S=length(M);
for gmmiteration=1:1:100
for w=1:1:5
WORD=DATA(:,(w-1)*F+1:1:(w-1)*F+F);
for i=1:1:F
DEN=0;
for j=1:1:S
DEN=DEN+P(j)*gaussvalue(WORD(:,i),M{j},COV{j});
end
for k=1:1:S
A{w}(i,k)=P(k)*gaussvalue(WORD(:,i),M{k},COV{k})/(DEN+eps);
end
end
end

for k=1:1:S
DEN=0;
for t=1:1:F
for x=1:1:5
DEN=DEN+A{x}(t,k);
end
end
for x=1:1:5
    for t=1:1:F
m(x,k,t)=A{x}(t,k)/(DEN+eps);
    end
end
end
for k=1:1:S
MEAN{k}=0;
COV{k}=0;
for x=1:1:5
    WORD=DATA(:,(x-1)*F+1:1:(x-1)*F+F);
    for t=1:1:F
    MEAN{k}=MEAN{k}+m(x,k,t)*WORD(:,t);
    end
```

```
end
for x=1:1:5
    WORD=DATA(:,(x-1)*F+1:1:(x-1)*F+F);
    for t=1:1:F
    COV{k}=COV{k}+m(x,k,t)*(WORD(:,t)-MEAN{k})*(WORD(:,t)-MEAN{k})';
    end
end
end
DEN=0;
for k=1:1:S
    for x=1:1:5
        for t=1:1:F
DEN=DEN+A{x}(t,k);
        end
    end
end
for k=1:1:S
    NUM=0;
for x=1:1:5
    for t=1:1:F
        NUM=NUM+A{x}(t,k);
    end
end
P(k)=(NUM)/(DEN+eps);
end
end
GMMMODEL{word}{1}=P;
GMMMODEL{word}{2}=COV;
GMMMODEL{word}{3}=M;
end
save GMMMODEL mx E

%gmmtestingforspeech.m
clear all
close all
load GMM
load TESTLPC
s=[];
for i=1:1:3
    s=[s size(TESTLPC{i},1)];
end
DATA=[];
for i=1:1:3
    DATA=[DATA TESTLPC{i}(1:1:mx,:)];
end
DATA1=[];
for i=1:1:size(DATA,2)
    temp=DATA(:,i);
DATA1=[DATA1 reshape(temp,13,14)];
end
DATA1=E'*DATA1;
%Generating probability of the first word
S=5;
F=14;
COL=[];
for x=1:1:15
prob=[];
DATA=DATA1(:,(x-1)*F+1:1:(x-1)*F+F);
for y=1:1:3
prob=[prob genprob(S,F,DATA,GMMMODEL{y})];
[P,Q]=max(prob);
end
COL=[COL Q];
end
```

1.6 Unsupervised Learning System

The learning algorithms such as BPN, SVM, GMM, and HMM come under the category of the supervised learning algorithm. In this, the class label (target output, i.e., word) of every training data (e.g., speech segment) is known. In this, the learning system is trained to minimize the error between the target outputs and the actually obtained outputs. There are circumstances where the target output is not known; still, we need to model the learning system to classify the training sets in the finite number of groups. This is known as unsupervised learning algorithm.

1.6.1 Need for Unsupervised Learning System

1. Based on the discussion in Sect. 1.5, we need to classify all the frames into finite number of groups to initialize the unknown parameters.
2. To narrow down the search algorithm for the huge database. Suppose we would like to design the practical system that classifies large number of words (say $N > 1,000$). It is very difficult to model (say BPNN) the system to classify one among the N. Instead, we perform the unsupervised clustering algorithm to cluster the N words into 10 groups based on the best match (e.g., measured using the Euclidean distance). Each cluster now has a fewer number of words. Formulating the BPNN for the individual cluster becomes easier. Let BPNN_r be the BPNN model corresponding to the rth cluster.
3. During testing, the cluster number corresponding to the test word is first identified. Further the actual word is identified using the BPNN_r model (refer Fig. 1.15).

1.6.2 K-Means Algorithm

Given the n number of training vectors v_i with $i = 1, \ldots, n$ and the number of clusters r, assignment of the individual vector to the corresponding cluster using k-means algorithm is obtained as follows:

1. Initialize the centroids of r clusters, namely $c_1, c_2, \ldots c_r$.
2. Compute the Euclidean distance between the ith vector and jth centroid as d_{ij}.
3. Assign the ith vector to the $\arg(\min \sum_{j=1}^{j=r} d_{ij})$ cluster.
4. Compute the mean of all the vectors assigned to the ith cluster to obtain ith centroid c_i.
5. Steps 3–5 are repeated for finite number of iterations, and finally, obtained centroid vectors are stored in the database.
6. Given the test vector x to be classified among one of the r clusters. Compute the distance between the xth vector and the centroid of the jth, which is represented

as d_{xj}. Declare the test vector belongs to the arg(min $\sum_{j=1}^{j=r} d_{xj}$) cluster. This is known as nearest mean (refer Fig. 1.17) (NM) classifier.

7. Given the test vector x, compute the distance between the vector x and the nearest vector in jth cluster. Let it be d_{xj}. Declare the test vector belongs to arg(min $\sum_{j=1}^{j=r} d_{xj}$) cluster. This is known as nearest neighbor (refer Fig. 1.17) (NN) classifier with one neighbor.
8. The another way of executing NN classifier is as follows. Calculate the number of vectors belonging to jth cluster within the radius R surrounding the test vector x. Let it be n_{xj}. Declare the test vector belongs to the arg(max $\sum_{j=1}^{j=r} n_{xj}$) cluster. This is known as NN classifier (refer Fig. 1.17) with radius R.

1.6.3 Illustration of k-Means Algorithm

The number of states in GMM as the generative model of the frames is obtained using k-means algorithm. This also helps to initialize the mean vector and the covariance matrix of the individual state of the GMM. The training LPC frames collected from three speech segments are subjected to PCA for dimensionality reduction and are subjected to k-means algorithm. The total number of frames is equal to the total number of vectors that are subjected to k-means clustering. The number of states and the number of iterations are initialized as 5 and 100, respectively. The sum-squared Euclidean distances between the individual vectors (after dimensionality reduction) and the centroid of the corresponding cluster (*state* in case of GMM) are plotted (refer Fig. 1.16) against the iteration number. The original 3D scatter plot (obtained using the PCA-based dimensionality reduction technique of the LPC data) and the clusters obtained using the k-means algorithm are illustrated in Fig. 1.18.

1.6.4 Fuzzy k-Means Algorithm

In k-means algorithm, every vector is assumed to be belong to one of the r clusters. In fuzzy k-means algorithm, every vector is assumed to be member in all the clusters with different membership values. Let the membership value of the ith vector belonging to the jth cluster be represented as m_{ij}. The membership values are chosen such that J in (1.50) is maximized and satisfying (1.51) as described below.

$$J = \sum_{j=1}^{j=r} \sum_{i=1}^{i=N} m_{ij}^{p} d_{ij} \tag{1.50}$$

where r is the number of clusters, N is the number of vectors, p is the constant > 1, and d_{ij} is the Euclidean distance between the ith vector and the jth cluster. It is also noted that the membership value m_{ij} is larger if $\frac{\sum_{l=1}^{l=r} d_{il}}{d_{ij}}$ is smaller and vice versa (refer Fig. 1.19).

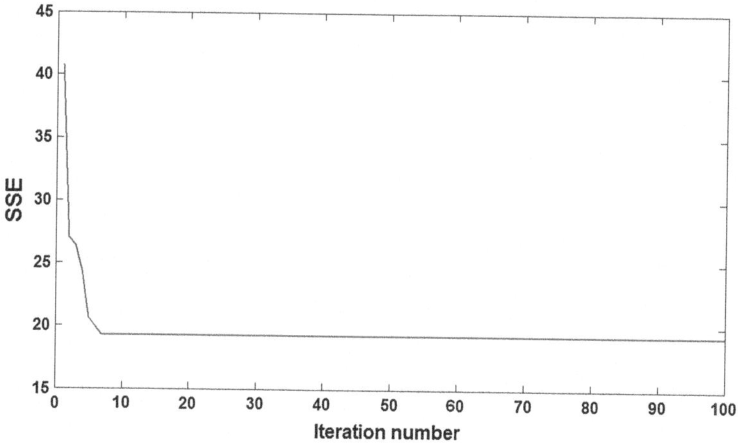

Fig. 1.16 Convergence graph obtained for k-means algorithm. It is noted that the significant reduction in SSE is achieved within the first 10 iterations

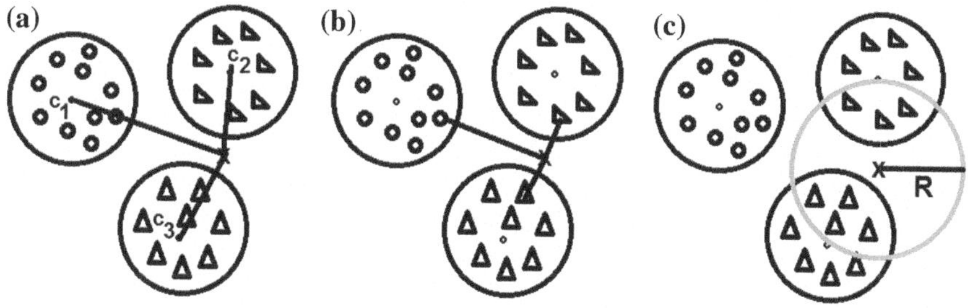

Fig. 1.17 Illustration of **a** nearest mean classifier, **b** nearest neighbor classifier with one neighbor, and **c** nearest neighbor classifier with radius R. X is the test vector

$$\sum_{j=1}^{j=r} m_{ij} = 1 \tag{1.51}$$

The Euclidean distance between the ith vector and the jth cluster is represented as d_{ij}. The membership values are calculated as follows:

1. If the Euclidean distance $d_{ij} < \eta$,

$$m_{ij} = 1,\ m_{il} = 0\,\forall l \neq j \tag{1.52}$$

2. If the Euclidean distance $d_{ij} \geq \eta$, m_{ij} is calculated as

$$m_{ij} = \frac{1}{d_{ij}^{\frac{1}{(p-1)}} \sum_{l=1}^{l=r} \frac{1}{d_{il}}^{\frac{1}{(p-1)}}} \tag{1.53}$$

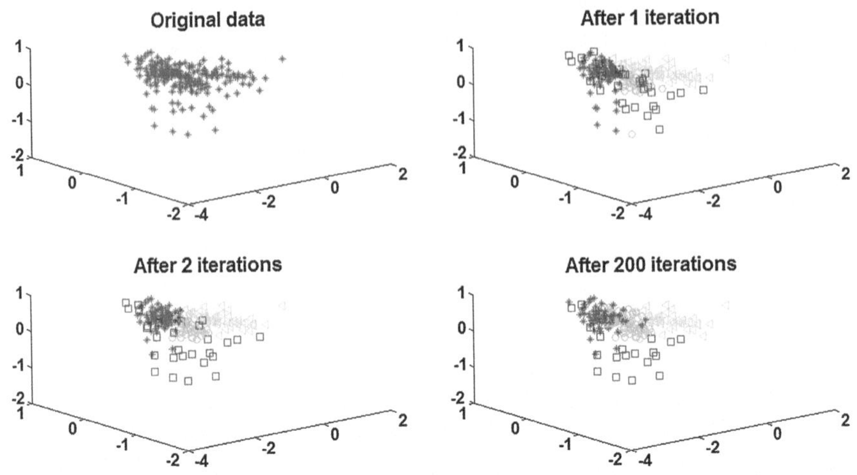

Fig. 1.18 Scatter plot of the 3D data before and after k-means clustering

$$\Rightarrow m_{ij} = \frac{1}{1 + \sum_{l=1,l\neq j}^{l=r} (\frac{d_{ij}}{d_{il}})^{\frac{1}{(p-1)}}} \tag{1.54}$$

3. The value of the fuzzy tuning parameter $m > 1$ is arbitrarily chosen.
4. The centroid of the jth cluster is computed as

$$c_j = \frac{\sum_{i=1}^{i=N} m_{ij}^p v_i}{\sum_{i=1}^{i=N} m_{ij}^p} \tag{1.55}$$

1.6.5 Steps Involved in Fuzzy k-Means Clustering

1. Initialize the centroids, iter $= 0$, $J(1) = 1000$ and t_h.
2. Increment iter $=$ iter $+ 1$. Compute the membership values using (1.52) and (1.54).
3. Compute the new set of centroids using (1.55).
4. Compute $J(\text{iter} + 1)$ using (1.50).
5. Repeat steps 2–3 until $(J(\text{iter}+1) - J(\text{iter}))^2$ is less than the threshold value t_h.

1.6.6 Illustration of Fuzzy k-Means Algorithm

The original 3D scatter plot (obtained using the PCA-based dimensionality reduction technique of the LPC data) and the clusters obtained using the fuzzy k-means algorithm are illustrated in Figs. 1.20, 1.21. The corresponding convergence graph is illustrated in Fig. 1.20. The number of clusters is chosen as four. The values of the p and η are chosen as 10 and 0.01, respectively.

```
%fuzzykmeansforspeech.m
function [M,COV,INDEX,m,D,J]=fuzzykmeansforspeech(DATA,n,N)
subplot(2,2,1)
plot3(DATA(1,:),DATA(2,:),DATA(3,:),'r*')
s1=size(DATA,2);
s2=size(DATA,2)/n;
r=0;
M=[];
J=[];
for i=1:1:n
M{i}=mean(DATA(:,(i-1)*s2+1:1:(i-1)*s2+s2)')';
end
p=10; th=0.01;
for iteration=1:1:N
%Computation of distance matrix
for i=1:1:s1
for j=1:1:n
    D(i,j)=(DATA(:,i)-M{j})'*(DATA(:,i)-M{j});
end
end
m=zeros(s1,n);
[P,Q]=find(D<th);
for i=1:1:length(P)
    m(P(i),:)=0;
    m(P(i),Q(i))=1;
end
%Computation of membership matrix
for i=1:1:s1
    for j=1:1:n
        s=0;
    for k=1:1:n
        if(k~=j)
       s=s+(D(i,j)/(D(i,k)+eps))^(1/(p-1));
        else
       s=s+1;
        end
    end
m(i,j)=1/(s+eps) ;
    end
end
[P,Q]=find(D<th);
for i=1:1:length(P)
    m(P(i),:)=0;
    m(P(i),Q(i))=1;
end

for j=1:1:n
NUM=0;
DEN=0;
    for i=1:1:s1
NUM=NUM+(m(i,j)^p)*DATA(:,i);
DEN=DEN+m(i,j)^p;
    end
    M{j}=NUM/(DEN+eps);
end
J=[J sum(sum((m.^p).*D))];
for k=1:1:n
    C{k}=[];
    INDEX{k}=[];
end
S=0;
```

```
for i=1:1:size(DATA,2)
    D1=[];
   for j=1:1:n
    D1=[D1 sum((DATA(:,i)-M{j}).^2)];
   end
[P,Q]=min(D1);
C{Q}=[C{Q} DATA(:,i)];
INDEX{Q}=[INDEX{Q} i];
end
end

for i=1:1:n
    COV{i}=cov(C{i}');
end
```

1.6.7 Kohonen Self-Organizing Map

To classify the given set of vectors into finite number of groups, the Kohonen self-organizing map (KSOM) model of the ANN is used. This is also called as vector quantization. In case of BPNN, for every input vector, the corresponding target vector is known. The weight matrices are adjusted to obtain the target vector for the corresponding input vector. But in case of KSOM, the output vector is not known. For every training input vector, the particular neuron in the output layer is obtained as the winner neuron and is declared as the group number. Hence, clustering is achieved using KSOM.

KSOM consists of two layers, namely input layer and output layer (refer Fig. 1.22). Every neuron in the input layer is connected with every neuron in the hidden layer. Let m and n be the number of neurons in the input layer and the output layer, respectively. The weight connecting the ith neuron in the input layer with the jth neuron in the output layer is represented as w_{ij}. The column vector $W_j = [w_{1j}\ w_{2j}\ w_{3j}\ \ldots w_{mj}]^T$ is the weight vector assigned to the jth neuron of the output layer. This is called as the jth weight vector. Let X_k be the kth column input vector, and let $d(X_k, W_j)$ be the squared Euclidean distance between the kth input vector and the jth weight vector W_j. The jth neuron is treated as the winner neuron for the kth input vector X_k and is obtained as $j = \arg(\min_{l=1,\ldots,n}(d(X_k - W_l)))$. For the given set of input vectors, obtaining the weight matrix that minimizes the sum-squared euclidean

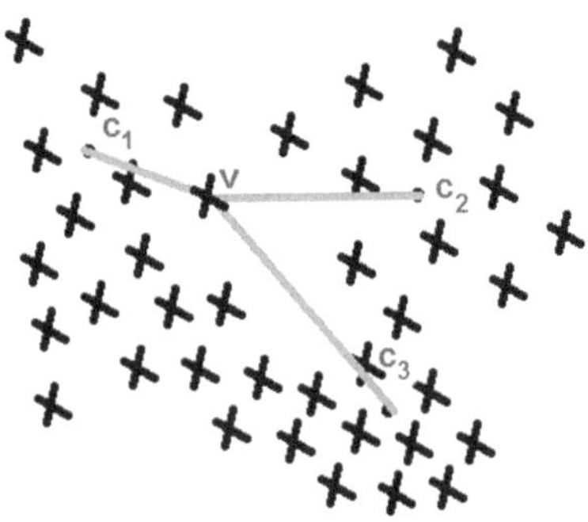

Fig. 1.19 Illustration of fuzzy k-means algorithm. The membership value of the arbitrary vector v with the first cluster is larger than that of the arbitrary vector v with the second and the third clusters

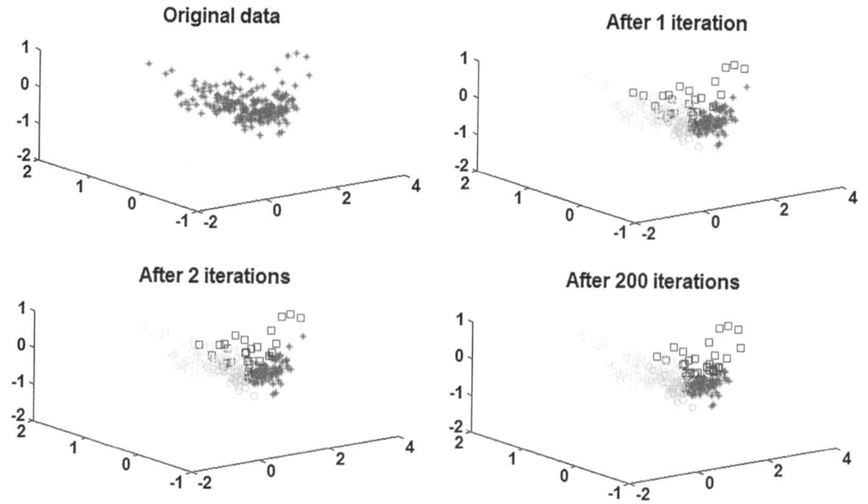

Fig. 1.20 Scatter plot of the 3D data before and after fuzzy k-means clustering

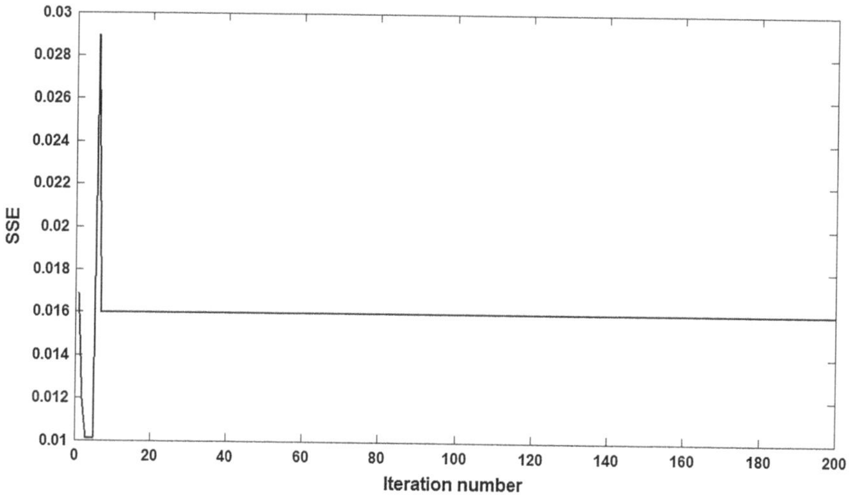

Fig. 1.21 Convergence graph obtained for the fuzzy k-means algorithm. The overshoot at the beginning is due to (1.53)

distances between all the input vectors and the corresponding winner weight vectors is known as training phase. The steps involved in the training phase are summarized below.

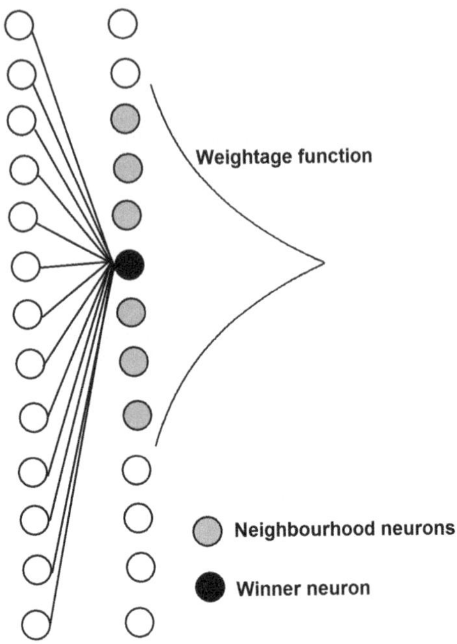

Fig. 1.22 Illustration of KSOM ANN

1.6.7.1 Training Phase for KSOM

1. Randomly initialize the weight matrix of size $m \times n$ with uniform distribution over the range -1 to 1. Note that the neurons in the output layer represent the index of the cluster.
2. The input vectors are normalized to the range -1 to 1. Initialize iteration number iter $= 1$ and number of iterations I.
3. For the first input vector X_1, identify the $j = \arg(\min_{l1,\ldots,n}(d(X_k - W_l)))$. j is treated as the winner neuron.
4. Choose the neighborhood neurons near to the winner neurons $l = j-k, j-k+1, \ldots, j+k$, where k is the integer constant.
5. Adjust the weight vectors corresponding to the neighborhood neurons such that the winner neuron is given the largest weightage. Also, more weightages are given for the neurons that are near to the winner neuron and lesser weightages are given to the neurons that are far away from the winner neuron. This helps to strengthen the role of the winner neuron corresponding to the input vector X_1. The factor $G_l = \exp^{-c(j-l)^2}$, where c is the constant used to calculate the weightage of the lth neuron. Note that weightage and the weight vector are different.
6. Thus, the weight vector W_l is updated as follows:

$$W_l(\text{iter}+1) = W_l(\text{iter}) + \eta_{\text{iter}}(X_1 - W_l(\text{iter})) \tag{1.56}$$

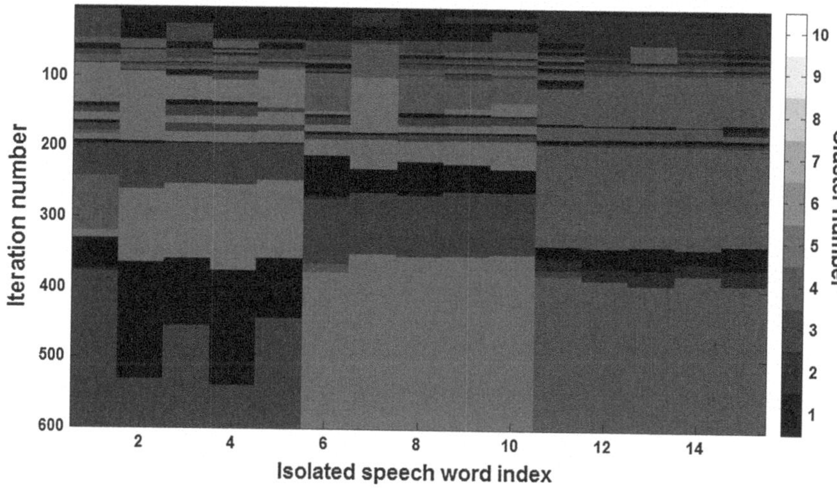

Fig. 1.23 Illustration of clustering using KSOM

where η_{iter} is the learning rate in the iter-th iteration and $l = j - k, \ldots, j + k$.

7. Steps 3–6 are repeated for every input vectors X_i for $i = 1, \ldots, N$, where N is the number of training data. This is one iteration.
8. Learning rate η, constant c, and the neighborhood size k are decreased once in every iteration.
9. The weight vectors are updated for I number of iterations, and the final weight vectors completely describe the trained KSOM network.

The winner neuron j corresponding to the arbitrary test vector Z is obtained using the trained KSOM network, and test vector Z is declared as the one belonging to the jth cluster.

1.6.8 Illustration of KSOM

The linear predictive coefficients are collected from three distinct words, and the PCA-based dimensionality reduction (refer Sect. 1.7.1) is performed to obtain the data set. The number of samples collected for every word is 10. 50 % of the collected data are used to train KSOM, and the remaining data are used for testing. The percentage of success obtained is 100 % (15 out of 15). The winner neurons are obtained as 3, 5, and 4, respectively, for the first, second, and third classes (three distinct words). Clustering achieved using the KSOM algorithm is displayed in Fig. 1.23. It is noted after 40th iteration that first five data (corresponding to the word 1) are clustered as 3rd group. The next five data (corresponding to the word 2) are clustered as 5th group, and the last five data are clustered as 4th group.

```
%ksomforspeech.m
clear all
close all
load TRAINLPC
s=[];
for i=1:1:3
    s=[s size(TRAINLPC{i},1)];
end
mx=min(s);
DATA=[];
for i=1:1:3
    DATA=[DATA TRAINLPC{i}(1:1:mx,:)];
end
[E,D]=eigs(cov(DATA'),14);
DATA=E'*DATA;
ma=max(max(abs(DATA)));
DATA=DATA/ma;
N=10;
W=rand(size(DATA,1),N);
k=2;
c=0.1;
eta=0.1;
for iteration1=1:1:20
eta=eta*0.1;
    for iteration2=1:1:200
for i=1:1:size(DATA,2);
D=sum((repmat(DATA(:,i),1,N)-W).^2);
[P,Q]=min(D);
l=Q-k:1:Q+k;
[P1,Q1]=find(l>0);
index=l(Q1);
G=exp(-c*(k-index).^2);
for j=1:1:length(index)
W(:,j)=W(:,j)+eta*G(j)*(DATA(:,i)-W(:,j) );
end
end
    end
end
%Verification using training data
COLTRAIN=[];
for i=1:1:size(DATA,2);
D=sum((repmat(DATA(:,i),1,N)-W).^2);
[P,Q]=min(D);
COLTRAIN=[COLTRAIN Q];
end

load TESTLPC
s=[];
for i=1:1:3
    s=[s size(TESTLPC{i},1)];
end
DATA=[];
for i=1:1:3
    DATA=[DATA TESTLPC{i}(1:1:mx,:)];
end
DATA=E'*DATA;
DATA=DATA/ma;
COLTEST=[];
for i=1:1:size(DATA,2);
D=sum((repmat(DATA(:,i),1,N)-W).^2);
[P,Q]=min(D);
COLTEST=[COLTEST Q];
end
```

1.7 Dimensionality Reduction Techniques

The feature vectors collected from the speech signal for isolated speech classification and speaker classification are large. If the techniques such as BPNN, SVM, unsupervised learning techniques such as k-means algorithm, fuzzy k-means algorithm, and KSOM are used, the computational complexity and the memory requirement have been increased. Hence, the vectors are mapped from the feature dimensional space to the lower-dimensional space. This is known as dimensionality reduction techniques. The most commonly used dimensionality reduction techniques are principal component analysis (PCA), linear discriminant analysis (LDA), kernel LDA (KLDA), and independent component analysis (ICA).

1.7.1 Principal Component Analysis

Let vectors in the feature dimensional space be represented as $v_i, \ \forall \ i = 1, \ldots, N$. The size of any vector v_i is $M \times 1$. The vector is mapped to the lower-dimensional space using the transformation matrix W^T. The size of the matrix W is $M \times N$, where $N << M$. We need to identify the matrix to map the vector v_i in the feature dimensional space to the corresponding vector u_i in the lower-dimensional space. The set of vectors u_i and v_i are related as $u_i = W^T v_i$. Let the scatter matrix computed in the feature dimensional vectors be represented as $S_v = \sum_{i=1}^{i=N} (v_i - \mu)^T (v_i - \mu)$, where $\mu = \frac{\sum_{i=1}^{i=N} v_i}{N}$ is the mean vector computed in the feature dimensional space. Let the scatter matrix computed using the lower-dimensional vectors be computed as S_u and be related to S_v as $S_u = W^T S_v W$.

It is noted that the diagonal elements of the scatter matrix S_v give the scaled version of the variances of the individual elements of the feature dimensional vector (i.e., $S_v(k, k) = N\sigma_{v_k}^2$). Similarly, $S_u(k, k) = N\sigma_{u_k}^2$. They are related as $S_u(k, k) = W_k^T S_v W_k$, where W_k is the kth column of the matrix W. The transformation matrix W is optimized such that the variances of the individual elements of the transformed vector are maximized. This is achieved as follows.

Maximize $W_k^T S_v W_k$ with the constraint $W_k^T W_k = 1$. This is solved using Lagrangian method. The Lagrangian equation for the optimization problem is given as

$$L(W_k, \lambda_k) = W_k^T S_v W_k - \lambda_k (W_k^T W_k - 1) \tag{1.57}$$

Differentiating (1.56) and equate to zero to get the optimal value as $S_v W_k = \lambda_k W_k$. As S_v is the positive semi-definite matrix, the eigenvectors of the matrix S_v are always non-negative. Also,

$$W_k^T S_v W_k = W_k^T \lambda_k W_k = \lambda_k W_k^T W_k = \lambda_k$$

Hence, λ_k is identified as the actual maximized scaled (N times) variance of the kth element of the lower-dimensional vector u. Hence, the eigenvector corresponding

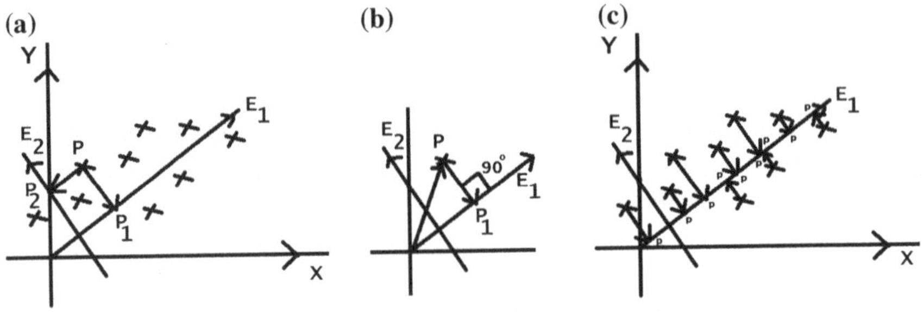

Fig. 1.24 Illustration of PCA using 2D to 1D conversion. P_1 is the 1D projection on the eigenvector corresponding to the 2D point P on the XY plane. The 2D vectors are represented as X

to the largest eigenvalue maximizes $W_k^T S_v W_k$. Thus, the eigenvectors corresponding to the N significant eigenvalues are arranged column-wise to form the transformation matrix W.

1.7.2 Illustration of PCA Using 2D to 1D Conversion

Consider that the vectors are scattered in the XY plane (refer Fig. 1.24a). The eigenvectors (E_1, E_2) are identified in the direction such that the variances of the projected points on the corresponding eigenvectors are maximum. Let the arbitrary point P in the XY plane be represented with respect to the XY basis as $[x_1, y_1]^T$. The projection of the arbitrary point P on the eigenvector E_1 with respect to the E_1 axis is the point P_1. So the vector joining the origin and the point P_1 is represented as $P_1 E_1$. As the projected vector is computed as $P - P_1 E_1$ and is perpendicular to the eigenvector E_1 (refer Fig. 1.24b), we get $(P - P_1 E_1)^T E_1 = 0$. This implies that $P_1 = E_1^T P$. Note that the $E_1^T E_1 = 1$. Thus, the projected point on the E_1 axis with respect to the $E_1 E_2$ basis is represented as $P_1 = E_1^T P$. Similarly, the projected point on the E_2 with respect to the $E_1 E_2$ basis is represented as $P_1 = E_2^T P$. Note that the variance of the projected vectors (points) on the E_1 axis is maximum (refer Fig. 1.24c as expected).

1.7.3 Illustration of PCA

The number of LPC coefficients obtained from the particular speech data is 182. They are obtained from 30 speech data. The significant eigenvalues are plotted in Fig. 1.25. Note that the number of significant eigenvalues is 29. Hence, 29 PCA bases are computed. But for illustration, 2D and 3D projected data are shown in Fig. 1.26.

```
%pcaforspeech.m
function [PCA]=pcaforspeech(DATA)
%The vectors are arranged in the columnwise in the matrix DATA
[E,D]=eig(cov(DATA'));
[P,Q]=find(diag(D)>0.001);
PCA=E(:,P(length(P)):-1:P(1));
```

1.7.4 Linear Discriminant Analysis

The PCA technique does not care about the discrimination information of the individual clusters. Hence, the variances of the individual elements of the projected vector belonging to different clusters are also maximized when PCA is used. To circumvent this, linear discriminant analysis (LDA) is formulated. In this technique, the variances of the individual elements of the projected centroid vectors are maximized. Simultaneously, the variances of the individual elements of the projected vectors corresponding to the identical clusters are minimized. Hence, LDA helps in bringing down the vectors closer to each other and simultaneously separating the vectors farther from each other in the projected lower-dimensional space. The LDA consists of two positive definite scatter matrices, namely between-class scatter matrix (S_B) and within-class scatter matrix (S_W) defined as follows. Let v_{ij} be the ith vector corresponding to the jth cluster. Also, let c_j be the centroid of the jth cluster, n_j be the number of vectors in the jth cluster, c be the overall centroid of the data, and r be the total number of clusters.

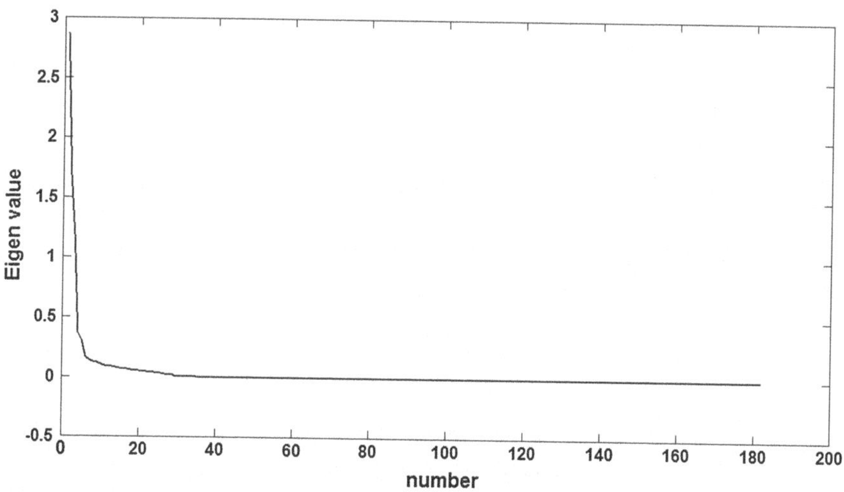

Fig. 1.25 Eigenvalues of the scatter matrix computed using 30 speech data (182 dimensional LPC)

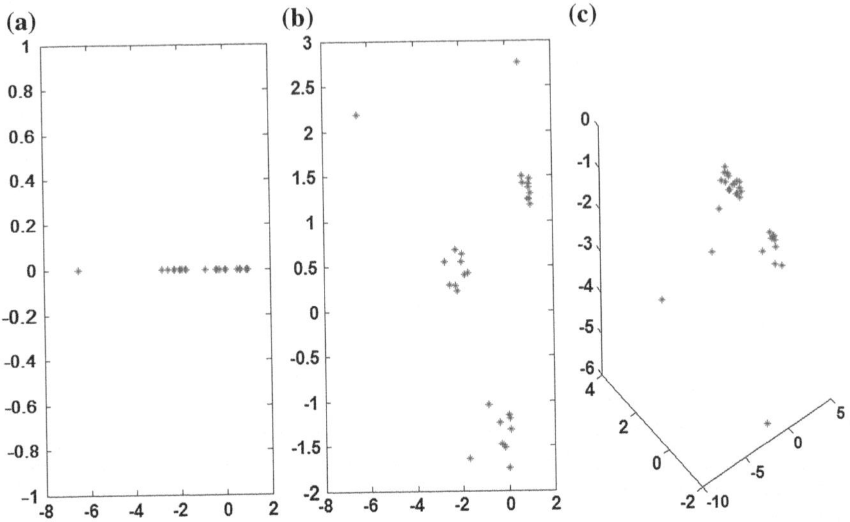

Fig. 1.26 Dimensionality reduction of the 30 speech data (182 dimensional LPC). **a** Dimensionality reduction to 1D, **b** dimensionality reduction to 2D, and **c** dimensionality reduction to 3D

$$S_W = \sum_{j=1}^{j=r} \sum_{i=1}^{i=n_j} [v_{ij} - c_j]^T [v_{ij} - c_j] \tag{1.58}$$

$$S_B = \sum_{j=1}^{j=r} n_j [c_j - c]^T [c_j - c] \tag{1.59}$$

It is noted that the diagonal elements of the matrix $W^T S_W W$ measure the closeness of the vectors belonging to the identical cluster in the projected lower-dimensional space. Similarly, the diagonal elements of the matrix $W^T S_B W$ measure the separation of the centroid vectors of various clusters in the projected lower-dimensional space. Thus, the transformation matrix is obtained such that

$$\frac{\mathrm{trace}(W^T S_B W)}{\mathrm{trace}(W^T S_W W)} \tag{1.60}$$

is maximized. Consider that the ith column vector of the matrix W is represented as W_i. To maximize (1.60), the ratio of every (i, i) element of the matrix $W^T S_B W$ to the (i, i) element of the matrix $W^T S_B W$ is maximized, i.e., $\frac{W_i^T S_B W_i}{W_i^T S_W W_i}$ is maximized. This is solved by maximizing $W_i^T S_B W_i$ with the constraint $W_i^T S_W W_i = W_i^T S_{WW}^T S_{WW} W_i$ (i.e., $(S_{WW} W_i)^T (S_{WW} W_i) = 1$). It is noted that the matrix S_W is represented as $S_{WW}^T S_{WW}$ as S_W is the positive semi-definite matrix.
The objective function is maximized using the Lagrangian technique as follows.

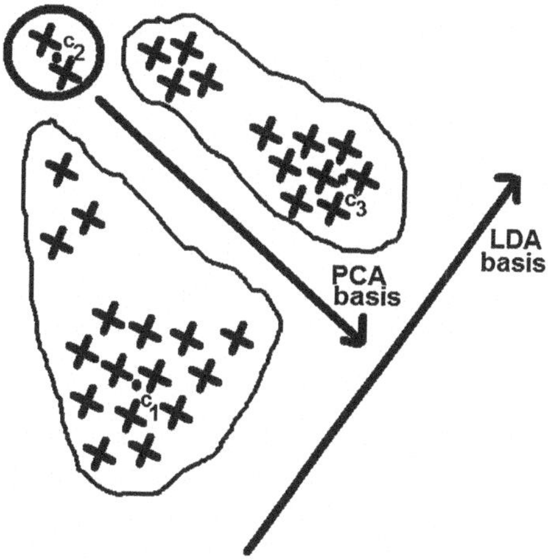

Fig. 1.27 Illustration of comparison between LDA basis and PCA basis for 2D point to 1D point conversion. Note that the LDA basis directs in the direction of the maximum variance of the centroids of the individual clusters

The Lagrangian equation is formulated as

$$J = W_i^T S_B W_i + \lambda_i((S_{WW} W_i)^T (S_{WW} W_i) - 1) \tag{1.61}$$

Differentiating (1.61) with respect to W_i and equate to zero, we get the following.

$$S_B W_i + \lambda_i S_W W_i \tag{1.62}$$

Thus, W_i is the eigenvector of the matrix $S_W^{-1} S_B$ that maximizes the objective function. Hence, the transformation matrix using LDA is obtained by column-wise arranging the eigenvectors of the matrix $S_W^{-1} S_B$ corresponding to the significant eigenvalues. Intuitively, we understand that the significant eigenvectors are in the direction of maximum variance of the centroid vectors (refer Fig. 1.27).

1.7.5 Small Sample Size Problem in LDA

To obtain the LDA basis, S_W must be the invertible matrix. If the number of attributes (elements) of the feature dimensional vector is greater than the number of feature vectors used to compute S_W, it becomes non-invertible matrix. This is known as small sample size problem.

1. The column space of the matrix S_W is the space spanned by the vectors used to compute S_W.
2. The size of the matrix S_W is $M \times M$. Let the number of vectors used to compute S_W be $\sum_{i=1}^{i=r} n_i \ll M$, and hence, the rank of the matrix S_W is at the most equal to $\sum_{i=1}^{i=r} n_i$, and thus, S_W is non-invertible.

3. There are many techniques proposed in the literature to circumvent the small sample size problems such as null-space projection, exponential LDA.

1.7.6 Null-Space LDA

1. Compute the within-class scatter matrix S_W. Project the vectors to the null space of the S_W. The arbitrary vector v is projected to the null space of the scatter matrix S_W to obtain v_n as follows:
   - Compute the eigenvectors corresponding to the significant eigenvalues of the matrix S_W. The vectors are arranged column-wise to obtain the matrix W.
   - The vector v is projected to the column space of the scatter matrix S_W as $v_c = WW^T v$.
   - The vectors v_c and v_n form the direct sum to form the vector v. Hence, $v_n = v - v_c$.
2. The between-class scatter matrix S_{BN} is computed using the projected vectors in the null space of S_W. The significant eigenvectors of the matrix S_{BN} form the LDA bases.

1.7.7 Kernel LDA

The vectors in the feature dimensional space are mapped to the lower-dimensional space using LDA. By intuition, we understand that if the vectors in the feature dimensional space are mapped to the HDS, the better separation between the clusters is achieved, and hence, LDA is applied effectively in the HDS to map the vectors to the lower-dimensional space as described below.

The ith vector in the jth cluster in the feature dimensional space is represented as v_{ij} $\forall\, i = 1, \ldots, n_i$ and $j = 1, \ldots, r$ are mapped to the HDS using the transformation ϕ to obtain the corresponding transformed vectors as $\phi(v_{ij})$ $\forall\, i = 1, \ldots, n_i$ and $j = 1, \ldots, r$. The LDA is formulated in the higher-dimensional space, i.e., the within-class and the between-class scatter matrices are computed in the HDS, which are represented as S_W^{ϕ} and S_B^{ϕ}, respectively. The LDA bases computed in the HDS satisfy

$$S_B^{\phi}\phi(W_k) = \lambda_k^{\phi} S_W^{\phi}\phi(W_k) \tag{1.63}$$

Note that $\phi(W_k)$ is the kth significant eigenvector in the HDS corresponding to the eigenvalue λ_k^{ϕ} in the HDS. The eigenvector $\phi\,(W_k)$ lies in the space spanned by the columns of the matrix $S_W^{-1}S_B$. This is in further lies in the space spanned by the vectors used to compute $S_W^{-1}S_B$ (i.e., $\phi(v_{ij})$ $\forall\, i = 1, \ldots, n_i$ and $j = 1, \ldots, r$). Hence, the eigenvector $\phi(W_k)$ is represented as $\phi(W_k) = M_{\phi}u_{\phi}$ for some arbitrary vector u_{ϕ}, where M_ϕ is given as follows. $M_\phi = [\phi(v_{11})\phi(v_{21})\ldots\phi(v_{n_1 1})\ldots\phi(v_{12})\phi(v_{22})\ldots\phi(v_{n_2 2})\ldots\phi(v_{n_r} r)]$. Hence, (1.63) is rewritten as follows.

$$S_B^\phi M_\phi u_\phi = \lambda_k^\phi S_W^\phi M_\phi u_\phi \tag{1.64}$$

Multiplying M_ϕ^T on both sides of (1.64), we get

$$M_\phi^T S_B^\phi M_\phi u_\phi = \lambda_k^\phi M_\phi^T S_W^\phi M_\phi u_\phi \tag{1.65}$$

The Gram matrix G_ϕ is computed as $G_\phi = M_\phi^T M_\phi$. It is noted that the matrix $M_\phi^T S_B^\phi M_\phi$ is the between-class scatter matrix computed using the corresponding column vectors of the Gram matrix G_ϕ, i.e., the first n_1 columns of the matrix G_ϕ are treated as the one belong to the first cluster. The next n_2 columns of the matrix G_ϕ belong to the second cluster and so on. Let it be represented as S_{BG}^ϕ. Similarly, $M_\phi^T S_W^\phi M_\phi$ is the within-class scatter matrix computed using the corresponding column vectors of the Gram matrix G_ϕ. Let it be represented as S_{WG}^ϕ. Rewriting (1.65) using S_{BG}^ϕ and S_{WG}^ϕ, we get the following.

$$S_{\mathrm{BG}}^\phi u_\phi = \lambda_k^\phi S_{\mathrm{WG}}^\phi u_\phi \tag{1.66}$$

It is noted that the vector u_ϕ is the kth significant eigenvector of the matrix $(S_{\mathrm{WG}}^\phi)^{-1} S_{\mathrm{BG}}^\phi$ in the HDS.

1.7.8 Kernel-Trick to Execute LDA in the Higher-Dimensional Space

1. The (i, j)th element of the matrix G_ϕ is computed as $\phi(v_{pq})^T \phi(v_{rs})$ for some integers p, q, r, and s. If the kernel function is defined as $\mathrm{kernel}(v_{pq}, v_{rs}) = \phi(v_{pq})^T \phi(v_{rs})$, there is no need for explicit transformation ϕ to map the vectors from the feature dimensional space to the HDS. This is known as "kernel-trick."
2. Thus, the LDA bases u_ϕ in the HDS are computed as the significant eigenvector of the matrix $(S_{\mathrm{WG}}^\phi)^{-1} S_{\mathrm{BG}}^\phi$ without the actual mapping to the HDS using the kernel function.
3. Let the transformation matrix computed in the HDS be given as W_ϕ. To map the arbitrary vector in the feature dimensional space v_i to the lower-dimensional space, the vector is first mapped to the HDS as $\phi(v_i)$ and the vector is mapped to the lower-dimensional space as $W_\phi^T \phi(v_i)$, which is computed as follows:

$$W_\phi^T \phi(v_i) = [\phi(E_1)\phi(E_2)\ldots\phi(E_N)]^T \phi(v_i) \tag{1.67}$$

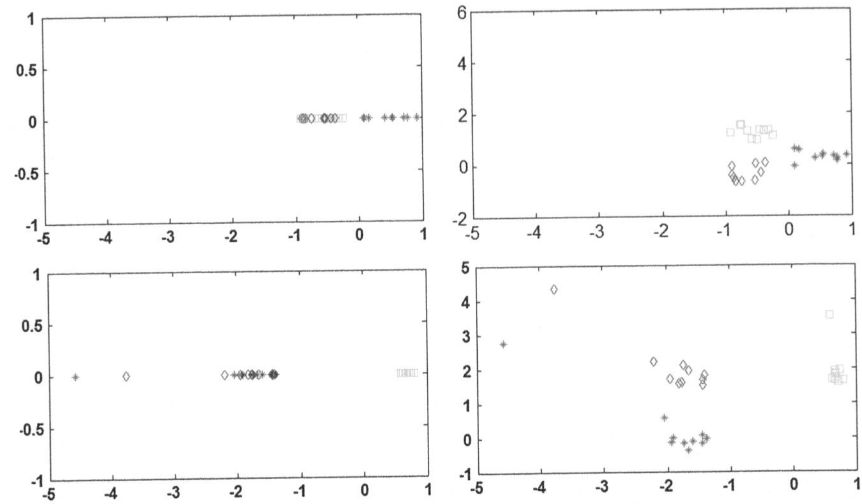

Fig. 1.28 Illustration of dimensionality reduction using LDA **a** 1D pseudo-inverse, **b** 2D pseudo-inverse, **c** 1D null-space LDA, and **d** 2D null-space LDA

$$\Rightarrow W_{\phi}^{T}\phi(v_i) = [u_{\phi 1}u_{\phi 2}\ldots u_{\phi N}]^{T}M_{\phi}^{T}\phi(v_i) \tag{1.68}$$

4. It is noted that $u_{\phi 1}, \ldots, u_{\phi N}$ are the N significant vectors of (1.65). Also note that $M_{\phi}^{T}\phi(v_i)$ is computed using the kernel function. Hence, the mapping from feature dimensional space to the lower-dimensional space without directly mapping to the intermediate HDS is achieved using the kernel function.
5. The mostly used kernel functions are listed in Table 1.1. It is also noted that the Gram matrix computed using the kernel function must be positive semi-definite matrix. This is known as Mercer's condition. This is needed to represent the Gram matrix as the product of $M^{T}M$. The list of properties used to construct the new kernel functions using the existing kernel functions is given in Table 1.2.

1.7.9 Illustration of Dimensionality Reduction Using LDA

30 LPC data corresponding to 3 distinct speech segments are subjected to LDA. The "small sample size" problem is taken care using pseudo-inverse technique and the null-space LDA technique. 1D and 2D projections are obtained using both the methods and are displayed in Fig. 1.28. The comparison of performance of various KLDA performed using various kernels are illustrated in Fig. 1.29 and Fig. 1.30.

```
%ldaforspeech
function [E,E2,DATAPI,DATANSP]=ldaforspeech(DATA)
%DATA is the cell, with each cell belongs to the individual cluster
%Computation of within-class scatter matrix
SW=0;
for i=1:1:3
    SW=SW+cov(DATA{i}');
end
MEAN=[];
for i=1:1:3
    MEAN=[MEAN mean(DATA{i}')'];
end
SB=cov(MEAN');
[E,D]=eig(pinv(SW)*SB);
D=D/max(max(D))
[P,Q]=find(D>0.01);
E=E(:,P(length(Q)):-1:P(1));
DATAPI=E'*cell2mat(DATA);
[E1,D1]=eig(SW);
[P1,Q1]=find(D>0.01);
DATA1=cell2mat(DATA);
DATAN=DATA1-E1(P1(length(Q)):-1:P1(1))*E1(P(length(Q)):-1:P1(1))'*DATA1;
M=[mean(DATAN(:,1:1:10)')'  mean(DATAN(:,11:1:20)')'  ...
mean(DATAN(:,1:21:30)')' ];
SBB=cov(M');
[E2,D2]=eig(SBB);
D2=D2/max(max(D2))
[P2,Q2]=find(D2>0.01)
E2=E2(:,P2(length(Q)):-1:P2(1));
DATANSP=E2'*cell2mat(DATA);

%kldaforspeech
function [E,E2,DATAPITR,DATANSPTR,DATAPITE,DATANSPTE]=kldaforspeech
(DATA1,DATA2,kerneltype)
%DATA1 and DATA2 are the cells, with each cell belongs to the individual cluster
%DATA1 belongs to the tranining data and DATA2 belongs to the testing data
%kerneltype
%1-Innerproduct: kerneltype
%2-Gaussian kerneltype
%3-Polynomial kerneltype
%4-Power exponential kerneltype
%5-Hyperbolic tangent
%6- Cauchy
%7-Inverse multi-quadric
DATA11=cell2mat(DATA1);
DATA22=cell2mat(DATA2);
switch kerneldata
    case 1
%Formation of training set
for i=1:1:size(DATA11,2)
    for j=1:1:size(DATA11,2)
TRDATA(j,i)=k00(DATA11(:,i),DATA11(:,j));
    end
end
%Formation of training set
for i=1:1:size(DATA11,2)
    for j=1:1:size(DATA11,2)
TEDATA(j,i)=k00(DATA22(:,i),DATA11(:,j));
    end
end
    case 2
%Formation of training set
for i=1:1:size(DATA11,2)
    for j=1:1:size(DATA11,2)
```

```
TRDATA(j,i)=k11(DATA11(:,i),DATA11(:,j),0.01);
    end
end
%Formation of testing set
for i=1:1:size(DATA11,2)
    for j=1:1:size(DATA11,2)
TEDATA(j,i)=k11(DATA22(:,i),DATA11(:,j),0.01);
    end
end
    case 3
%Formation of training set
for i=1:1:size(DATA11,2)
    for j=1:1:size(DATA11,2)
TRDATA(j,i)=k22(DATA11(:,i),DATA11(:,j),0.1,0.1);
    end
end
%Formation of testing set
for i=1:1:size(DATA11,2)
    for j=1:1:size(DATA11,2)
TEDATA(j,i)=k22(DATA22(:,i),DATA11(:,j),0.1,0.1);
    end
end
    case 4
%Formation of training set
for i=1:1:size(DATA11,2)
    for j=1:1:size(DATA11,2)
TRDATA(j,i)=k33(DATA11(:,i),DATA11(:,j),0.1,0.1);
    end
end
%Formation of testing set
for i=1:1:size(DATA11,2)
    for j=1:1:size(DATA11,2)
TEDATA(j,i)=k33(DATA22(:,i),DATA11(:,j),0.1,0.1);
    end
end
    case 5
%Formation of training set
for i=1:1:size(DATA11,2)
    for j=1:1:size(DATA11,2)
TRDATA(j,i)=k44(DATA11(:,i),DATA11(:,j),0.1,0.1);
    end
end
%Formation of testing set
for i=1:1:size(DATA11,2)
    for j=1:1:size(DATA11,2)
TEDATA(j,i)=k44(DATA22(:,i),DATA11(:,j),0.1,0.1);
    end
end
    case 6
%Formation of training set
for i=1:1:size(DATA11,2)
    for j=1:1:size(DATA11,2)
TRDATA(j,i)=k55(DATA11(:,i),DATA11(:,j),0.1);
    end
end
%Formation of testing set
for i=1:1:size(DATA11,2)
    for j=1:1:size(DATA11,2)
TEDATA(j,i)=k55(DATA22(:,i),DATA11(:,j),0.1);
    end
end
    case 7
%Formation of training set
```

```
for i=1:1:size(DATA11,2)
    for j=1:1:size(DATA11,2)
TRDATA(j,i)=k66(DATA11(:,i),DATA11(:,j),0.1);
    end
end
%Formation of testing set
for i=1:1:size(DATA11,2)
    for j=1:1:size(DATA11,2)
TEDATA(j,i)=k66(DATA22(:,i),DATA11(:,j),0.1);
    end
end
end
for i=1:1:3
DTR{i}=TRDATA(:,(i-1)*5+1:1:(i-1)*5+5);
end
for i=1:1:3
DTE{i}=TRDATA(:,(i-1)*5+1:1:(i-1)*5+5);
end
[E,E2,DATAPITR,DATANSPTR]=ldaforspeech1(DTR);
DATAPITE=E'*cell2mat(DTE);
DATANSPTE=E2'*cell2mat(DTE);

%k00.m
function [res]=k00(x1,x2)
x1=x1/sqrt(sum(x1.^2));
x2=x2/sqrt(sum(x2.^2));
res=x1'*x2;

%k11.m
function [res]=k11(x1,x2,c1)
x1=x1/sqrt(sum(x1.^2));
x2=x2/sqrt(sum(x2.^2));
c1=c1/2;
res=exp(-sum((x1-x2).^2)/(c1^2));

%k22.m
function [res]=k22(x1,x2,c2,c3)
x1=x1/sqrt(sum(x1.^2));
x2=x2/sqrt(sum(x2.^2));
c3=c3/2;
M=round(c2*8)+2;
res=((x1'*x2)/2+c3)^(M);

%k33.m
function [res]=k33(x1,x2,c4,c5)
x1=x1/sqrt(sum(x1.^2));
x2=x2/sqrt(sum(x2.^2));
res=exp(-sum((x1-x2).^2)/(c4^2))^(c5);

%k44.m
function [res]=k44(x1,x2,c6,c7)
x1=x1/sqrt(sum(x1.^2));
x2=x2/sqrt(sum(x2.^2));
res=tanh(c6*(x1'*x2)+c7);

%k55.m
function [res]=k55(x1,x2,c8)
x1=x1/sqrt(sum(x1.^2));
x2=x2/sqrt(sum(x2.^2));
res=1+(sum((x1-x2).^2)/c8);
res=1/res;

%k66.m
function [res]=k66(x1,x2,c9)
x1=x1/sqrt(sum(x1.^2));
x2=x2/sqrt(sum(x2.^2));
res=sqrt(sum((x1-x2).^2)+c9^2);
res=1/res;
```

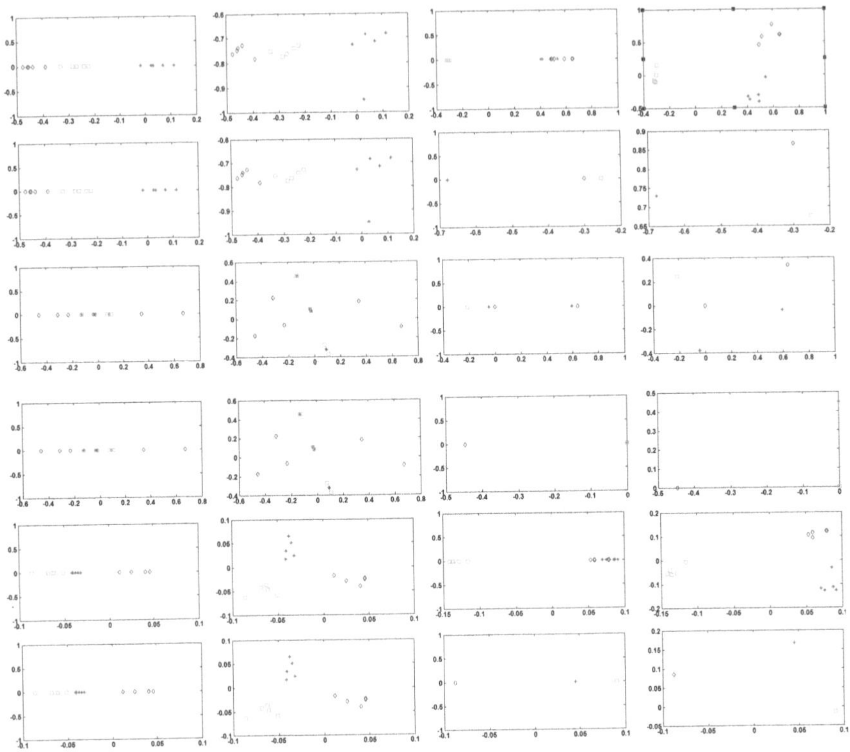

Fig. 1.29 Illustration of dimensionality reduction using KLDA. Column-wise: (1) 1D pseudo-inverse, (2) 2D pseudo-inverse, (3) 1D null-space LDA, and (4) 2D null-space LDA. Row-wise: (1–2) Projected (training–testing) data using inner product kernel, (3–4) Gaussian kernel, and (5–6) polynomial kernel type

1.8 Independent Component Analysis

In independent component analysis (ICA), the elements of the bases vectors are assumed to be independent to each other. In PCA, the covariance matrix computed using the PCA bases is the diagonal matrix. This indicates that the elements of the bases vectors of the PCA bases are uncorrelated with each other. This does not guarantee the independence of the individual elements. Given the training feature vectors, identifying the n bases whose elements are independent to each other is known as independent component analysis (ICA). Let n independent source signals be linearly mixed to obtain the m mixed signals. Let n independent source signals be represented as the random vector (refer Fig. 1.31) as $\overline{X} = [X_1\ X_2\ X_3 \ldots X_n]^T$. They are mixed using the matrix W sized $n \times n$ to obtain another set of random variables and are represented as the random vector as $[\overline{Y} = Y_1\ Y_2\ Y_3 \cdots Y_n]^T$. They are related as $\overline{Y} = W^T\overline{X}$. The training feature vectors are assumed to be the mixed signals and

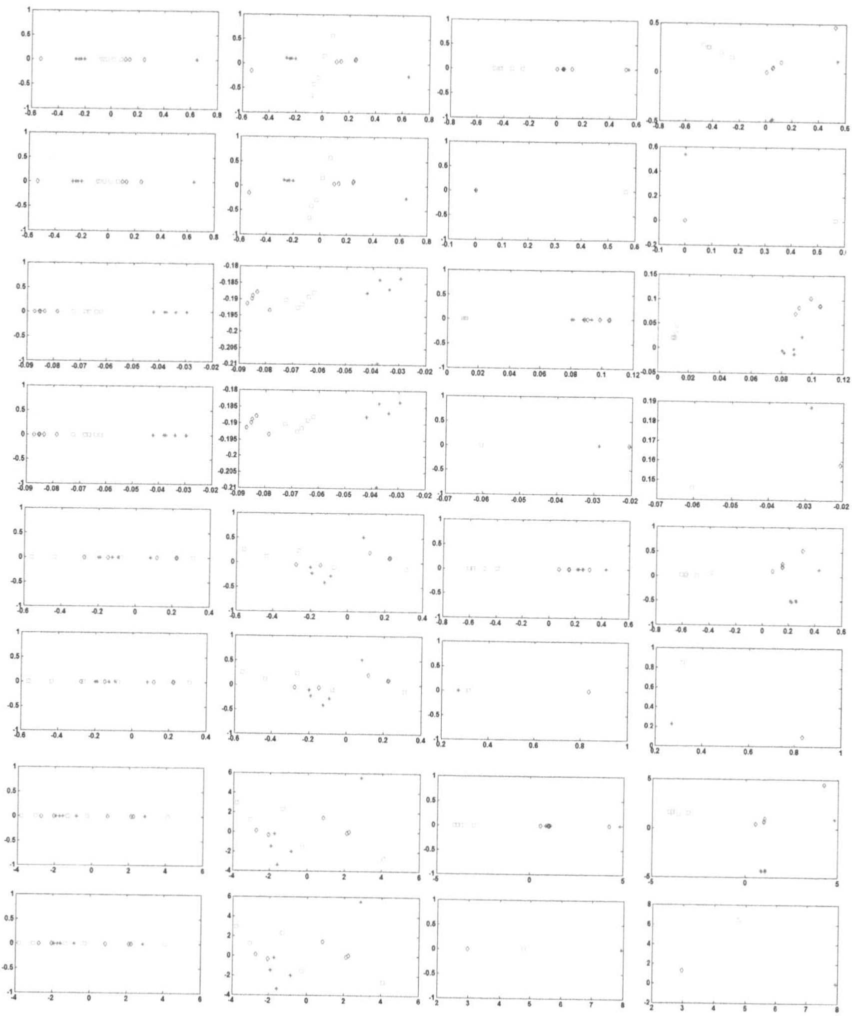

Fig. 1.30 Illustration of dimensionality reduction using KLDA. Column-wise: (1) 1D pseudo-inverse, (2) 2D pseudo-inverse, (3) 1D null-space LDA, and (4) 2D null-space LDA. Row-wise: (1–2) Projected (training–testing) data using power exponential, (3–4) hyperbolic tangent, (5–6) Cauchy, and (7–8) inverse multiquadric

are described by the random vector $\overline{Y}$. ICA involves identifying the transformation matrix W and the independent source signals described by the random vector $\overline{X}$.

Based on the central limit theorem, we understand that the random variables of arbitrary probability density functions or probability mass functions are linearly mixed, and we obtain the random variable that is more Gaussian (i.e., when the number of independent signals reaches infinity, the pdf of the mixed signals becomes Gaussian). We assume that the training feature vectors are the mixed signals and the

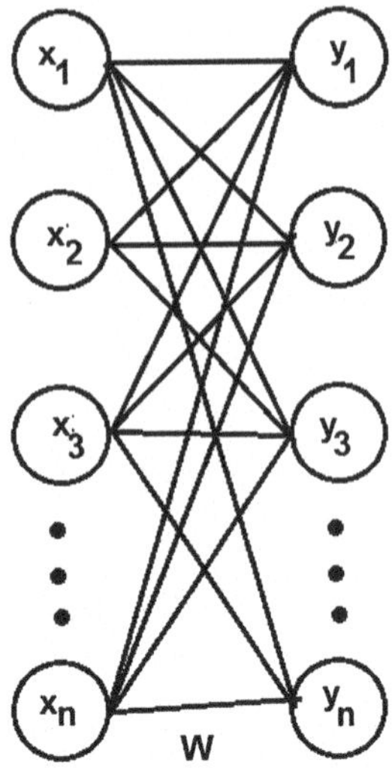

Fig. 1.31 Illustration of ICA

individual elements of the feature vectors are Gaussian. The individual elements of the independent source signals (bases that generate the feature vector space) are less Gaussian.

The Gaussianity of the random variable R with mean zero is measured using the kurtosis k as described below.

$$K(R) = E(R^4) - 3(E(R^2))^2 \tag{1.69}$$

If the random variable R having mean zero is Gaussian, then $K(R) = 0$. The ICA problem is formulated and solved using the kurtosis as described below.

1.8.1 *Solving ICA Bases Using Kurtosis Measurement*

1. The individual elements of the independent signals are less Gaussian and hence the absolute value of the kurtosis measured for the individual elements of the independent signals are maximum.
2. The variance of independent signals is assumed to be unity. This is achieved by satisfying $WW^T = 1$ (i.e., the rows of the matrix W are orthonormal to each other).

$$\overline{Y} = W^T\overline{X} \tag{1.70}$$

$$\Rightarrow \overline{X} = W\overline{Y} \tag{1.71}$$

$$J = \sum_{i=1}^{i=n} \text{abs}(K(X_i)) = \sum_{i=1}^{i=n} \text{abs}(E(X_i^4) - 3(E(X_i^2))^2) \tag{1.72}$$

$$= \sum_{i=1}^{i=n} \text{abs}(E(X_i^4) - 3) \tag{1.73}$$

3. As $E(X_i^4)$ is always positive, maximizing $\text{abs}(E(X_i^4) - 3)$ is equivalent to maximizing $E(X_i^4) - 3$.
4. Thus, ICA involves solving the transformation matrix W such that $J = \sum_{i=1}^{i=n} E(X_i^4) - 3$ is maximized. Note that the condition $W^T W = 1$ is used (i.e., the rows of the matrix W are orthonormal to each other).
5. To maximize J, the Lagrangian equation is formulated as follows:

$$L = \sum_{i=1}^{i=n} E\left(\left(\sum_{j=1}^{j=n} W_{ij} Y_j\right)^4\right) - 3 + \lambda(W_i^T W_i - 1) \tag{1.74}$$

where W_{ij} is the (i, j)th element of the matrix W and W_i is the ith row of the matrix W.
6. Differentiating (1.74) with respect to W_{ik} and equate to zero, we get the following:

$$4E((\sum_{j=1}^{j=n} W_{ij} Y_j)^3 Y_k) + 2\lambda(W_{ik}) = 0 \tag{1.75}$$

7. The iteration equation is formulated using (1.77) by choosing $\lambda = -2$ as

$$W_{ik}(t + 1) = E((\sum_{j=1}^{j=n} W_{ij}(t) Y_j)^3 Y_k) \tag{1.76}$$

where $W_i(t)$ is the W_i in the tth iteration.
8. The Lagrangian constraints do not care about the orthogonality condition, i.e., $W^T W = I$. It is noted as $C_Y = W^T C_X W$. If C_Y is the diagonal matrix, $W^T W = I$. Hence, the random vector $\overline{Y}$ (feature space) is preprocessed to obtain another random vector $\overline{Z}$ such that the covariance matrix is identity matrix. This is achieved using the transformation matrix $\overline{Z} = E^T D^{1/2}(\overline{Y} - \text{mean}(\overline{Y}))$, where E is the matrix with column vectors filled up with the eigenvectors of the covariance matrix C_Y and D is the diagonal matrix filled up with the eigenvalues in the diagonal. Thus, the procedure to obtain the ICA bases are summarized below.

1.8.2 Steps to Obtain the ICA Bases

1. Given the training set vectors $\overline{Y}$, obtain the transformed set of vectors using $\overline{Z}$ using the transformation matrix $\overline{Z} = E^T D^{1/2}(\overline{Y} - \text{mean}(\overline{Y}))$. This is used to obtain zero mean and unit covariance matrix of the random variable Z.
2. Randomly initialize the weight vectors W_{ik} $\forall\, i, k = 1, \ldots, n$ with magnitude 1. Also, rows of the matrix W are made orthonormal using Gram–Schmidt orthogonalization process.
3. Update W_{ik} using the following for finite number of iterations.

$$W_{ik}(t+1) = E(\sum_{j=1}^{j=n}(W_{ij}(t)Z_j)^3 Z_k) \tag{1.77}$$

4. Finally obtained W matrix is used to obtain the independent components $\overline{X}$ using $W\overline{Z}$. The rows of matrix W are explicitly orthogonalized using Gram–Schmidt orthogonalization procedure after every iteration.
5. It is noted that any mixing vector $\overline{Y}$ can be represented as the linear combinations of orthogonal column vectors of the matrix W. Hence, the column vectors of the matrix W form the ICA bases. A number of ICA bases are equal to the size of the mixing matrix $n \times n$.
6. Dimensionality reduction using ICA bases is obtained by first projecting the vectors to the lower-dimensional space using PCA. ICA bases are obtained using the projected space using the steps 2–5.

1.8.3 Illustration of Dimensionality Reduction Using ICA

Two experiments are performed. In the first experiment, two signals are generated. One is the sinusoidal with the particular frequency, and the another is the square wave. They are linearly mixed to obtain two mixed signals. ICA is applied to the mixed signals to obtain the independent signals. This is the illustration of the ICA to obtain the independent signals. The scatter plot of the original two signals, mixed signals, signals after PCA, and signals after ICA are displayed in Fig. 1.32, and the corresponding signals are plotted in Figs. 1.33, 1.34, 1.35, and 1.36. In the second experiment, the LPC speech signals are used. 15 LPC speech signals with 5 per distinct words are used as the training set to obtain the ICA basis. PCA is initially used to reduce the dimension of the signal from 182 to 14. Fourteen independent components are obtained using ICA. 2D scatter plot corresponding to the first two largest eigenvalues is plotted for the training set and the testing set as shown in Fig. 1.37.

```
%icabasisforspeech.m
function [W]=icabasisforspeech(DATA1)
temp=cell2mat(DATA1);
[E,D]=eig(cov(temp'));
E=E*D^(-1/2);
D1=D/max(max(D));
[P,Q]=find(D1>0.01);
E=E(:,P(length(P)):-1:P(1));
M=mean(temp')';
Y=real(E'*(temp-repmat(M,1,size(temp,2))));
CY=cov(Y');
check=1;
while(check==1)
W=rand(size(CY,1),size(CY,2));
if(rank(W)==size(CY,1))
W=gram(W')';
check=0;
end
end
for iteration=1:1:1
    for i=1:1:size(W,1)
        for j=1:1:size(W,2)
                    s1=0;
        for k=1:1:size(Y,2)
                        s=0;
    for l=1:1:size(Y,1)
        s=s+(W(i,l)*Y(l,k))
    end
    s1=s1+(s^3)*Y(j,k);
        end
        W(i,j)=s1/(size(Y,1)*size(Y,2));
        end
    end
W=gram(W')' ;
end


%gram.m
function [W]=gram(W)
for i=1:1:size(W,2)
    W(:,i)=W(:,i)/sqrt(sum(W(:,i).^2));
end
for i=1:1:size(W,2)-1
    s=W(:,i+1);
for j=1:1:i;
    s=s-(W(:,i+1)'*W(:,j))*W(:,j);
end
s=s/sqrt(sum(s.^2));
W(:,i+1)=s;
end
```

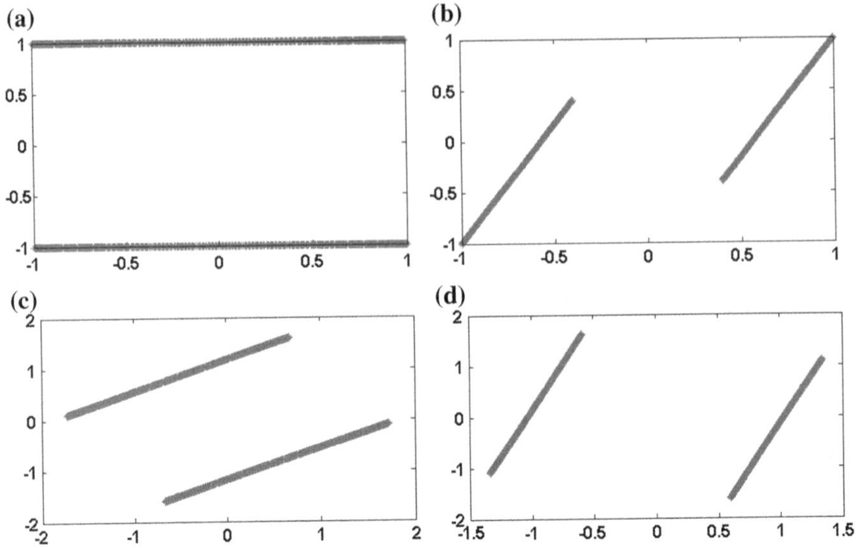

Fig. 1.32 Illustration of ICA using scatter plots of **a** original signals, **b** mixed signals, **c** uncorrelated signals using PCA, and **d** independent signals using ICA (after 1,000 iterations)

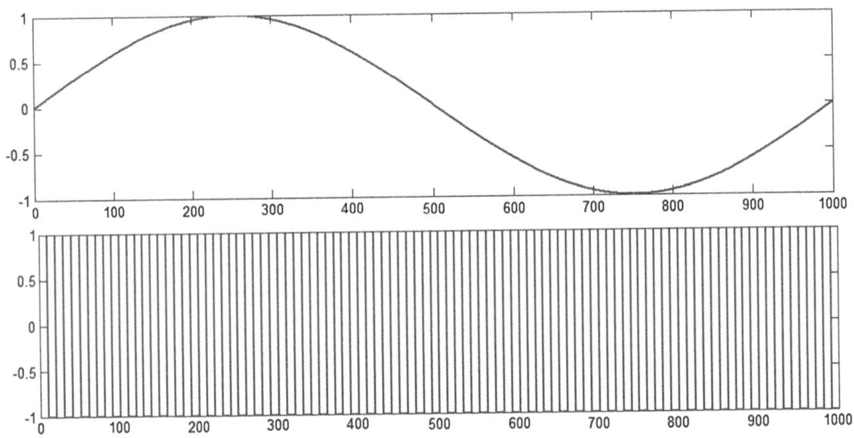

Fig. 1.33 Original signals

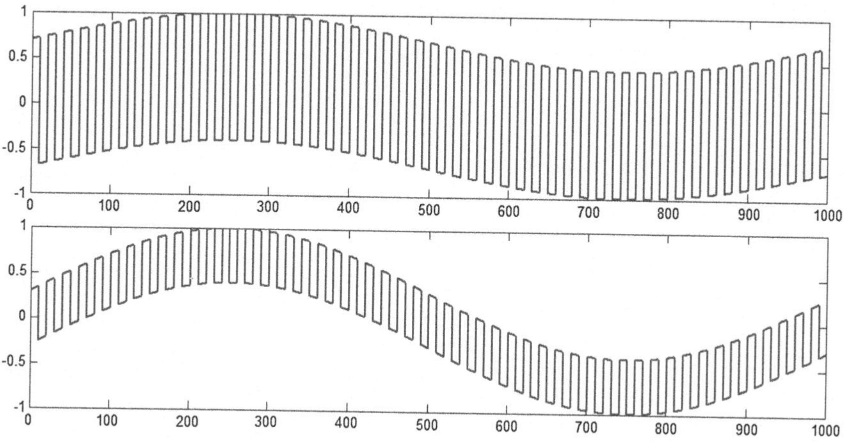

Fig. 1.34 Mixed signals

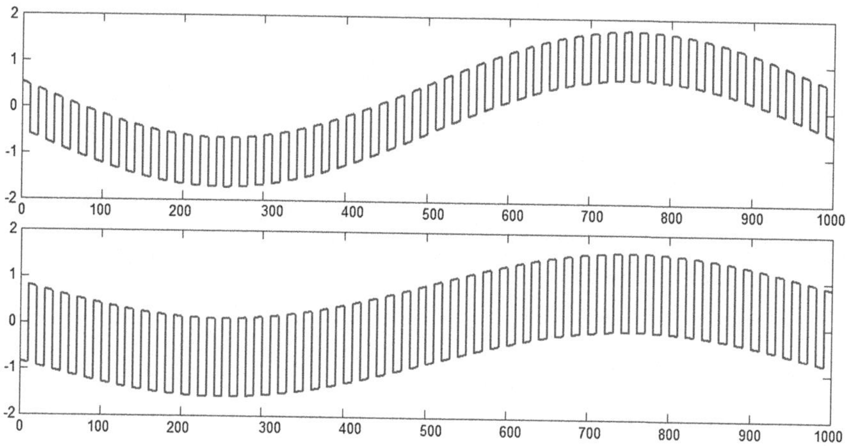

Fig. 1.35 Uncorrelated signals obtained using PCA

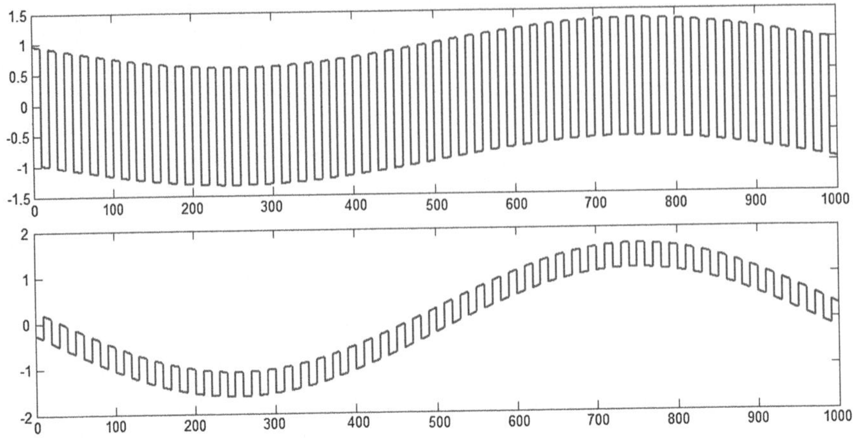

Fig. 1.36 Independent signals obtained using ICA

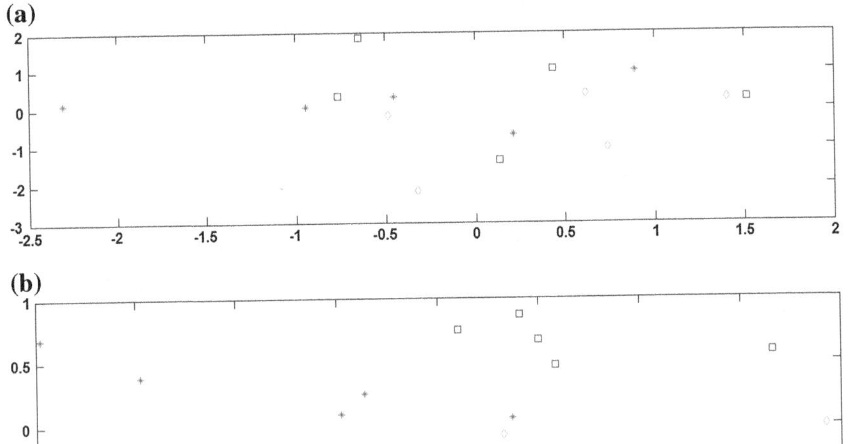

Fig. 1.37 **a** 2D scatter plot (corresponding to the largest two eigenvalues) using ICA for the training data and **b** 2D scatter plot using ICA for the testing data

Chapter 2
Speech Production Model

Abstract The continuous speech signal (air) that comes out of the mouth and the nose is converted into the electrical signal using the microphone. The electrical speech signal thus obtained is sampled to obtain the discrete signals and are stored in the digital system for further processing. This is digital speech processing. The speech signal model is broadly classified as the source-filter model and the probabilistic model. Source-filter model assumes the physical phenomenon for the production of speech signal. Probabilistic model like Hidden Markov Model (HMM), Gaussian Mixture Model (GMM) are the mathematical model that does not care about the physical phenomenon. Speech model is used to extract the feature vectors from the speech signal for isolated speech recognition and the speaker recognition. It is used to compress the speech signal for storage like in Code exited linear prediction (CELP). It is useful for converting text into speech, known as speech synthesis. It is also used for continuous speech recognition. This chapter deals with the source-filter model of speech production.

2.1 Introduction

The air that comes out of the lungs passes through the vocal tract and comes out of the mouth and the nose to obtain the continuous speech signal. The air coming out of lungs are either sent directly to the vocal tractor or altered using the vocal chord vibrations before sending to the vocal tract. The speech signals with vocal chord vibrations are known as voiced speech signals. The speech signals without the vocal chord vibrations are known as unvoiced speech signals. The velum is used to close the nose path,so that the speech signal is coming out only through the mouth. The vocal tract path is adjusted using tongue and velum to produce different speech signal. Thus lung, vocal chord, vocal tract, tongue, velum, mouth and nose are the integral part that produces the speech signal (refer Appendix F).

E. S. Gopi, *Digital Speech Processing Using Matlab*,
Signals and Communication Technology, DOI: 10.1007/978-81-322-1677-3_2,
© Springer India 2014

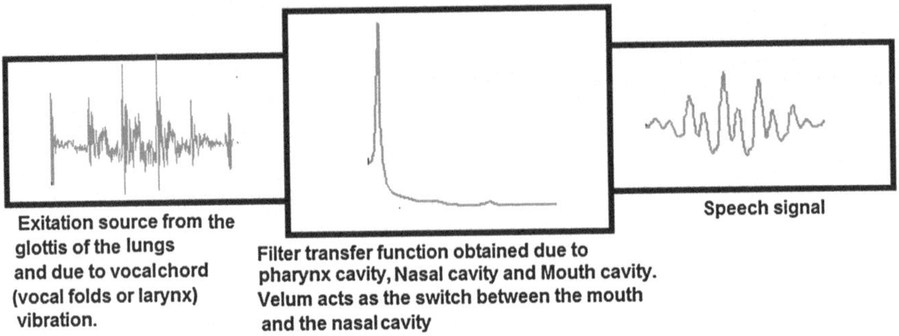

Fig. 2.1 Source-filter model of the speech production

2.2 1-D Sound Waves

The sound waves are longitudinal waves. It produces the disturbance along the direction of the flow (refer Fig. 2.1). The disturbance is in the form of compression and rarefaction. In source-filter model, the source is either the noise (air from the lungs) or the impulse stream (vocal chord vibration with the particular frequency) and the filter is the vocal-tract. The filter is assumed as the cascade connections of the tubes with different cross-sectional area. The length of the tube is usually less than the wavelength of the produced sound wave. Hence speech–sound waves are assumed to travel in one-dimensional direction. This model is known as 1-D sound wave.

2.2.1 *Physics on Sound Wave Travelling Through the Tube with Uniform Cross-Sectional Area A*

Consider the small segment of the tube (shaded region). When the sound wave crosses the small segment, the change in the physical entities (refer Table 2.1) like force (F), pressure (P), volume flow in terms of volume/s (S), velocity (V) are described below. Let the tube is kept along the direction of X-axis. Let the points I and J (refer Fig. 2.2) are at the distances x and $x + \Delta x$ from the origin. The pressure as point I is represented as P. Hence the pressure at J is computed as follows

Fig. 2.2 1-D sound wave travelling through the tube with uniform cross-sectional area A

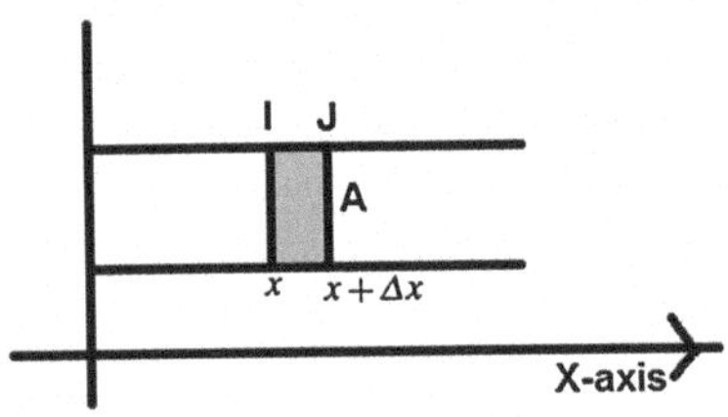

$$P + \frac{\partial P}{\partial x}\Delta x \tag{2.1}$$

The velocity is computed as the volume flow per unit area across the tube ($V = \frac{S}{A}$). Let the velocity at I is given as V. Hence the velocity at J is computed as follows

$$V + \frac{\partial S}{A\partial x}\Delta x \tag{2.2}$$

The volume of the air in the element is computed as $L = A\Delta x$. The rate of change of volume is computed as

$$\frac{\partial L}{\partial t} = \frac{A\partial x}{\partial t} = A\frac{\partial x}{\partial t} \tag{2.3}$$

Note that $\frac{\partial x}{\partial t}$ is the change in the velocity in the element. From (2.2) and (2.3), we get the following

$$\frac{\partial L}{\partial t} = A\frac{\partial S}{A\partial x}\Delta x = \frac{\partial S}{A\partial x}L \tag{2.4}$$

Net force (N_F) in the cross section is obtained as the difference between the force at I and at J.
Force at I is computed as pressure at A × cross sectional area = PA. Force at J is computed as pressure at J × cross sectional area = $(P + \frac{\partial P}{\partial x}\Delta x)A$. Thus the netforce in the cross section in the x-direction is given as follows.

$$N_F = PA - (P + \frac{\partial P}{\partial x}\Delta x)A = -\frac{\partial P}{\partial x}A\Delta x \tag{2.5}$$

Net force in the cross section is also computed as mass × accelaration. Also density of the air ρ × volume of the cross section gives the mass of the air inside the cross section. Recall $V = \frac{S}{A}$ and also note that acceleration is the rate of change of velocity and hence it is computed as $\frac{\partial V}{\partial t} = \frac{\partial S}{A\partial t}$. Hence the netforce is computed as follows

$$N_F = \rho L\frac{\partial S}{A\partial t} \tag{2.6}$$

Equating (2.5), (2.6) and $L = A\Delta x$, we get the following.

$$-\frac{\partial P}{\partial x}A\Delta x = \rho L\frac{\partial S}{A\partial t} \tag{2.7}$$

$$\Rightarrow -\frac{\partial P}{\partial x}A = \rho L\frac{\partial S}{A\Delta x\partial t} \tag{2.8}$$

$$\Rightarrow -\frac{\partial P}{\partial x}A = \rho\frac{\partial S}{\partial t} \tag{2.9}$$

From ideal gas law inside the cross section, we get, $PL = nRT$, where P is the pressure, L is the volume inside the cross section, $n = \frac{\text{mass}}{\text{molecular weight of the air}} = \frac{\text{volume} \times \text{density}}{\text{molecular weight of the air}} = \frac{L\rho}{M}$ is the number of moles, R is the gas constant, T is the temperature in kelvin. From the above discussion, we get the following.

$$PL = \frac{L\rho}{M} RT \Rightarrow P = \frac{\rho}{M} RT \tag{2.10}$$

The square of the speed of the sound depends only on temperature and is given as $c^2 = \gamma \frac{RT}{M}$. Hence,

$$P\gamma = \frac{\rho\gamma}{M} RT = \rho c^2 \tag{2.11}$$

The transfer of sound energy inside the cross section is faster so that we can assume that there is no transfer of heat energy and hence we assume it as the adiabatic process inside the cross section. Hence it obeys PL^γ = constant. This implies the following

$$PL^\gamma = \text{constant} \tag{2.12}$$

$$\Rightarrow \frac{\partial (PL^\gamma)}{\partial t} = 0 \tag{2.13}$$

$$\Rightarrow P\gamma L^{\gamma-1} \frac{\partial L}{\partial t} + L^\gamma \frac{\partial P}{\partial t} \tag{2.14}$$

$$\Rightarrow \frac{\partial L}{\partial t} \frac{P\gamma}{L} + \frac{\partial P}{\partial t} = 0 \tag{2.15}$$

From (2.4), (2.11) and (2.15), we get the following

$$\frac{\partial S}{A \partial x} L \frac{P\gamma}{L} + \frac{\partial P}{\partial t} = 0 \tag{2.16}$$

$$\Rightarrow \frac{\partial S}{A \partial x} P\gamma + \frac{\partial P}{\partial t} = 0 \tag{2.17}$$

$$\Rightarrow \rho c^2 \frac{\partial S}{\partial x} = -A \frac{\partial P}{\partial t} \tag{2.18}$$

The sound flow in the segment is described by (2.9) and (2.18).

2.2.2 Solution to (2.9) and (2.18)

Differentiating (2.9) with respect to t and (2.18) with respect to x, we get the following

$$-\frac{\partial P}{\partial x} A = \rho \frac{\partial S}{\partial t} \tag{2.19}$$

$$\Rightarrow -\frac{\partial^2 P}{\partial x \partial t}A = \rho\frac{\partial^2 S}{\partial t^2} \tag{2.20}$$

$$\rho c^2\frac{\partial S}{\partial x} = -A\frac{\partial P}{\partial t} \tag{2.21}$$

$$\Rightarrow \rho c^2\frac{\partial^2 S}{\partial x^2} = -A\frac{\partial^2 P}{\partial t \partial x} \tag{2.22}$$

Using (2.20) and (2.22), we get the following

$$\rho c^2\frac{\partial^2 S}{\partial x^2} = \rho\frac{\partial^2 S}{\partial t^2} \tag{2.23}$$

$$\Rightarrow \frac{\partial^2 S}{\partial t^2} = c^2\frac{\partial^2 S}{\partial x^2} \tag{2.24}$$

Let $u = x + ct$ and $v = x - ct$. $\frac{\partial^2 S}{\partial x^2}$ is computed in terms of u and v as follows

$$\frac{\partial S}{\partial t} = \frac{\partial S}{\partial u}\frac{\partial u}{\partial t} + \frac{\partial S}{\partial v}\frac{\partial v}{\partial t} \tag{2.25}$$

$$\Rightarrow \frac{\partial S}{\partial t} = \frac{\partial S}{\partial u}c + \frac{\partial S}{\partial v}(-c) \tag{2.26}$$

$$\Rightarrow \frac{\partial^2 S}{\partial t^2} = c(\frac{\partial^2 S}{\partial u^2}(c) + \frac{\partial^2 S}{\partial u \partial v}(-c) - \frac{\partial^2 S}{\partial v \partial u}(c) - \frac{\partial^2 S}{\partial v^2}(-c)) \tag{2.27}$$

$$\Rightarrow \frac{\partial^2 S}{\partial t^2} = c^2(\frac{\partial^2 S}{\partial u^2} + \frac{\partial^2 S}{\partial v^2} - 2\frac{\partial^2 S}{\partial u \partial v}) \tag{2.28}$$

Similarly $\frac{\partial^2 S}{\partial x^2}$ is computed as follows

$$\frac{\partial^2 S}{\partial x^2} = (\frac{\partial^2 S}{\partial u^2} + \frac{\partial^2 S}{\partial v^2} + 2\frac{\partial^2 S}{\partial u \partial v}) \tag{2.29}$$

Using (2.28) and (2.29), (2.24) is rewritten as follows

$$c^2(\frac{\partial^2 S}{\partial u^2} + \frac{\partial^2 S}{\partial v^2} + 2\frac{\partial^2 S}{\partial u \partial v}) = c^2(\frac{\partial^2 S}{\partial u^2} + \frac{\partial^2 S}{\partial v^2} - 2\frac{\partial^2 S}{\partial u \partial v}) \tag{2.30}$$

$$\Rightarrow \frac{\partial^2 S}{\partial u \partial v} = 0 \Rightarrow \frac{\partial S}{\partial u} = f(v) \tag{2.31}$$

$$\Rightarrow S(x,t) = g(v) + h(u) = g(x - ct) + h(x + ct) \tag{2.32}$$

Note that f, g and h are arbitrary functions. Represent $g(x - ct)$ as $S^+(t - \frac{x}{c})$ and $g(x + ct)$ as $-S^-(t + \frac{x}{c})$, we get the following

$$S(x,t) = S^+(t - \frac{x}{c}) - S^-(t + \frac{x}{c}) \tag{2.33}$$

Using (2.18) we get the pressure equation as follows

$$\frac{\partial S}{\partial x} = (-\frac{1}{c})(S'^{+}(t - \frac{x}{c}) + S'^{-}(t + \frac{x}{c})) \tag{2.34}$$

$$\Rightarrow \frac{\partial P}{\partial t} = c\rho(S'^{+}(t - \frac{x}{c}) + S'^{-}(t + \frac{x}{c})) \tag{2.35}$$

$$\Rightarrow P(x,t) = \frac{\rho c}{A}(S^{+}(t - \frac{x}{c}) + S^{-}(t + \frac{x}{c})) \tag{2.36}$$

Note that $S^{+}(t-\frac{x}{c})$ is the volume flow in the positive direction of x-axis and $S^{-}(t-\frac{x}{c})$ is the volume flow in the negative direction of x-axis. Thus the netflow in the positive direction is given as (2.33). Also the pressure at (t, x) is the constant times absolute sum of volume flow in both the directions as given in (2.36).

2.3 Vocal Tract Model as the Cascade Connections of Identical Length Tubes with Different Cross-Sections

Consider that the vocal tract is modelled as the cascade of three identical length (L) tubes with the cross sectional areas as A_1, A_2 and A_3 respectively (refer Fig. 2.3). The inlet volume flow in the forward direction of the ith tube is represented as P_i. The outlet volume flow in the forward direction of the ith tube is represented as R_i. Similarly the inlet and outlet volume flow in the reverse direction of the ith tube is represented as S_i and Q_i respectively. The relationship between P_i, Q_i, R_i, S_i are given as follows

$$P_i(t) = R_i(t + \tau) \tag{2.37}$$

$$\Rightarrow R_i(t) = P_i(t - \tau)S_i(t) = Q_i(t + \tau) \tag{2.38}$$

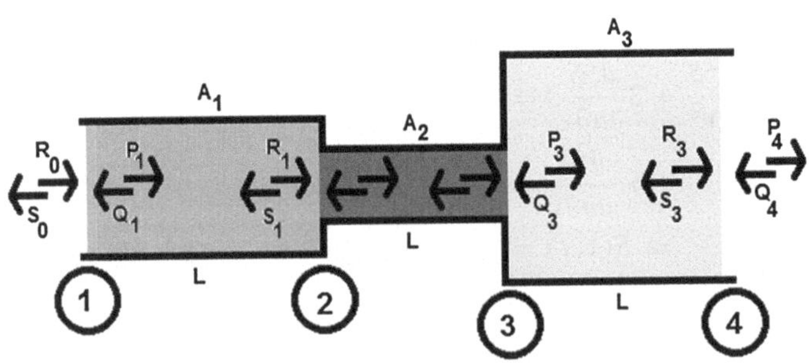

Fig. 2.3 Cascade of three tubes with different cross-sectional areas as the model of the vocal tract

where τ is the delay. The delay is the time required for the sound wave to travel through the single tube of length L, which is computed as $\tau = L/c$. If the signal is sampled with sampling time T_s, we get the following

$$R_i(nT_s) = P_i(nT_s - \tau) \quad (2.39)$$

$$S_i(nT_s) = Q_i(nT_s + \tau) \quad (2.40)$$

If $\tau = \frac{T_s}{2}$ and representing the (2.39) and (2.40) in discrete form, we get the following

$$R_i(n) = P_i(n - \frac{1}{2}) \quad (2.41)$$

$$S_i(n) = Q_i(n + \frac{1}{2}) \quad (2.42)$$

Representing in z-domain, we get the following

$$R_i(Z) = P_i(Z)Z^{\frac{1}{2}} \quad (2.43)$$

$$S_i(Z) = Q_i(Z)Z^{-\frac{1}{2}} \quad (2.44)$$

Let the input vector of the ith segment is represented as $I_{Si} = [P_i\ Q_{Si}]^T$ and the output vector of the segment is represented as $O_{Si} = [R_i\ S_i]^T$. They are related with the matrix $[M_{Si}]$ as $[O_{Si}] = [M_{Si}][I_{Si}]$, where M_{Si} is given as follows. Representing in the matrix form in z-domain, we get the following $\begin{bmatrix} Z^{\frac{1}{2}} & 0 \\ 0 & Z^{-\frac{1}{2}} \end{bmatrix}$. Let the input vector of the ith junction is represented as $I_{Ji} = [R_{i-1}\ S_{i-1}]^T$ and the output vector of the segment is represented as $O_{Ji} = [P_i\ Q_i]^T$ (refer Fig. 2.3). They are related using the matrix M_{Ji} as $[O_{Ji}] = [M_{Ji}][I_{Ji}]$. M_{Ji} is computed as follows. At the ith junction, there is the continuation in the pressure and the volume flow as mentioned below.

- Volume flow continuity

$$R_{i-1} - S_{i-1} = P_i - Q_i \quad (2.45)$$

- Pressure continuity

$$\frac{\rho c}{A_{i-1}}(R_{i-1} + S_{i-1}) = \frac{\rho c}{A_i}(P_i + Q_i) \quad (2.46)$$

$$\Rightarrow \frac{(R_{i-1} + S_{i-1})}{A_{i-1}} = \frac{(P_i + Q_i)}{A_i} \quad (2.47)$$

$$\Rightarrow (R_{i-1} + S_{i-1})(A_i) = (P_i + Q_i)(A_{i-1}) \quad (2.48)$$

Multiplying (2.45) with A_i we get the following

$$(R_{i-1} - S_{i-1})A_i = (P_i - Q_i)A_i \quad (2.49)$$

Adding (2.48) and (2.49), we get the following

$$2R_{i-1}A_i = P_i(A_i + A_{i-1}) + Q_i(A_i - A_{i-1}) \tag{2.50}$$

$$\Rightarrow R_{i-1} = P_i\frac{(A_i + A_{i-1})}{2A_i} + Q_i\frac{(A_i - A_{i-1})}{2A_i} \tag{2.51}$$

Subtracting (2.48) and (2.49), we get the following

$$2S_{i-1}A_i = P_i(A_{i-1} - A_i) + Q_i(A_{i-1} + A_i) \tag{2.52}$$

$$\Rightarrow S_{i-1} = P_i\frac{(A_{i-1} - A_i)}{2A_i} + Q_i\frac{(A_{i-1} + A_i)}{2A_i} \tag{2.53}$$

Let $r_i = \frac{A_i - A_{i-1}}{A_i + A_{i+1}}$ and hence

$$\frac{A_{i-1}}{A_i} = \frac{1-r_i}{1+r_i} \tag{2.54}$$

Using (2.54), we get the following

$$\frac{(A_i + A_{i-1})}{2A_i} = \frac{1}{2}(1 + \frac{A_{i-1}}{A_i}) = \frac{1}{1+r_i} \tag{2.55}$$

$$\frac{(A_{i-1} - A_i)}{2A_i} = \frac{1}{2}(\frac{A_{i-1}}{A_i} - 1) = \frac{-r_i}{1+r_i} \tag{2.56}$$

Thus R_{i-1} and S_{i-1} are expressed interms of r as follows

$$R_{i-1} = P_i\frac{1}{1+r_i} - Q_i\frac{-r_i}{1+r_i} \tag{2.57}$$

$$S_{i-1} = P_i\frac{-r_i}{1+r_i} + Q_i\frac{1}{1+r_i} \tag{2.58}$$

Thus the matrix M_{Ji} is given as $\frac{1}{1+r_i}\begin{bmatrix} 1 & -r_i \\ -r_i & 1 \end{bmatrix}$. It is noted that the matrix M_{Ji} is identical in z-domain also. The transfer function of the system is given as $\frac{P_4(Z)}{R_0(Z)}$. This is computed using the relationship between the vector $I_0(Z)$ and $O_4(Z)$ as $O_4(Z) = M_{J1}(Z)M_{S1}(Z)M_{J2}(Z)M_{S2}(Z)M_{J3}(Z)M_{S3}M_{J4}(Z)I_0(Z)$, which is computed as follows.

$$\begin{bmatrix} R_0(Z) \\ S_0(Z) \end{bmatrix} = \frac{1}{1+r_1}\begin{bmatrix} 1 & -r_1 \\ -r_1 & 1 \end{bmatrix}\begin{bmatrix} Z^{\frac{1}{2}} & 0 \\ 0 & Z^{-\frac{1}{2}} \end{bmatrix}\frac{1}{1+r_2}\begin{bmatrix} 1 & -r_2 \\ -r_2 & 1 \end{bmatrix}\begin{bmatrix} Z^{\frac{1}{2}} & 0 \\ 0 & Z^{-\frac{1}{2}} \end{bmatrix}$$
$$\frac{1}{1+r_3}\begin{bmatrix} 1 & -r_3 \\ -r_3 & 1 \end{bmatrix}\begin{bmatrix} Z^{\frac{1}{2}} & 0 \\ 0 & Z^{-\frac{1}{2}} \end{bmatrix}\frac{1}{1+r_4}\begin{bmatrix} 1 & -r_4 \\ -r_4 & 1 \end{bmatrix}\begin{bmatrix} P_4(Z) \\ Q_4(Z) \end{bmatrix}$$

Note that R_0 is coming from the lung openings and P_4 is coming out of mouth and nose. As there is no feedback in the mouth opening during speech, Q_4 is equated to zero and solving the $\frac{P_4(Z)}{R_0(Z)}$ gives the transfer function. Note that r_i is defined as the reflection co-efficient of ith segment. The values for r_i ranges from -1 to 1. It is also noted that A_0 is the area of the opening from the lungs to the vocal chord (glottis opening) and A_4 is assumed to be large finite value. On simplification, we get the following

$$\begin{bmatrix} R_0(Z) \\ S_0(Z) \end{bmatrix} = \frac{1}{(1+r_1)(1+r_2)(1+r_3)(1+r_4)} \begin{bmatrix} Z^{\frac{1}{2}} & -r_1 Z^{-\frac{1}{2}} \\ -r_1 Z^{\frac{1}{2}} & Z^{-\frac{1}{2}} \end{bmatrix} \begin{bmatrix} Z^{\frac{1}{2}} & -r_2 Z^{-\frac{1}{2}} \\ -r_2 Z^{\frac{1}{2}} & Z^{-\frac{1}{2}} \end{bmatrix}$$
$$\begin{bmatrix} Z^{\frac{1}{2}} & -r_3 Z^{-\frac{1}{2}} \\ -r_3 Z^{\frac{1}{2}} & Z^{-\frac{1}{2}} \end{bmatrix} \begin{bmatrix} Z^{\frac{1}{2}} & -r_4 Z^{-\frac{1}{2}} \\ -r_4 Z^{\frac{1}{2}} & Z^{-\frac{1}{2}} \end{bmatrix} \begin{bmatrix} Z^{-\frac{1}{2}} & 0 \\ 0 & Z^{\frac{1}{2}} \end{bmatrix} \begin{bmatrix} P_4(Z) \\ 0 \end{bmatrix}$$

On further simplification, we get the following

$$\begin{bmatrix} R_0(Z) \\ S_0(Z) \end{bmatrix} = \frac{1}{(1+r_1)(1+r_2)(1+r_3)(1+r_4)} \begin{bmatrix} Z + r_1 r_2 & -r_2 - r_1 Z^{-1} \\ -r_1 Z - r_2 & r_1 r_2 + Z^{-1} \end{bmatrix}$$
$$\begin{bmatrix} Z + r_3 r_4 & -r_3 - r_4 Z^{-1} \\ -r_3 Z - r_4 & r_3 r_4 + Z^{-1} \end{bmatrix} \begin{bmatrix} Z^{-\frac{1}{2}} & 0 \\ 0 & Z^{\frac{1}{2}} \end{bmatrix} \begin{bmatrix} P_4(Z) \\ 0 \end{bmatrix}$$

Thus the transfer function of the vocal tract is given as follows

$$\frac{P_4(Z)}{R_0(Z)} = \frac{Z^{-\frac{1}{2}}}{(Z + r_1 r_2)(Z + r_3 r_4) + (r_2 + r_1 Z^{-1})(r_4 + r_3 Z)} \quad (2.59)$$

$$= \frac{Z^{-\frac{1}{2}}}{Z^2 + (r_1 r_2 + r_3 r_4 + r_2 r_3)Z + r_1 r_2 r_3 r_4 + r_2 r_4 + r_1 r_3 + r_1 r_4 Z^{-1}} \quad (2.60)$$

$$= \frac{Z^{-\frac{3}{2}}}{1 + (r_1 r_2 + r_3 r_4 + r_2 r_3)Z^{-1} + +r_1 r_4 Z^{-2} + r_1 r_2 r_3 r_4 + r_2 r_4 + r_1 r_3} \quad (2.61)$$

Note that the factor $Z^{-\frac{3}{2}}$ is due to the delay introduced by the three segments. The transfer function of the vocal tract is identified as the ALL POLE third order filter. In general if vocal chord is assumed to have r segments, we get the transfer function becomes the rth order all pole filter with the delay factor of $Z^{-r/2}$. Thus the generalized transfer function of the rth order vocal tract filter is given as follows

$$V(Z) = \frac{Z^{-\frac{r}{2}}}{1 - \sum_{k=1}^{k=r} a_k Z^{-k}} \quad (2.62)$$

The length of the vocal tract is approximately 15 cm for adults. If the sampling frequency is $F_s = 8000\,Hz$, the length of the each segment is given as $\frac{c}{2F_s} = \frac{340}{16000} = 0.02125$ m. Hence number of segments are usually assumed as $\frac{0.15}{0.02125} \cong 7$. Hence the order of the filter is assumed around 7 for the sampling frequency of 8000 Hz.

2.4 Modelling the Vocal Tract from the Speech Signal

The sound wave that comes out from the lungs is the noise, which passes through the vocal tract filter to produce the particular speech signal. This type of speech signal is known as unvoiced speech signal. The sound wave that is produced by the vocal chord gets mixed with the wave that comes out of the lungs is passed through the vocal tract filter to produce the particular speech signal. This type of speech signal is known as unvoiced speech signal. In both the cases, the speech signal is modelled as the convolution of the sound source with the vocal tract filter. In Z-domain, speech signal is the product of the Z-transformation of the source signal with the Z-Transformation of the vocal tract. Let the source signal is represented as $I(Z)$ and output speech signal $S(Z)$ and are related as $\frac{S(Z)}{I(Z)} = V(Z) = \frac{Z^{-\frac{n}{2}}}{1-\sum_{k=1}^{k=n} a_k Z^{-k}}$. Rewriting the expression without delay we get the following.

$$\frac{S(Z)}{I(Z)} = \frac{1}{1-\sum_{k=1}^{k=n} a_k Z^{-k}} \Rightarrow S(n) = I(n) + \sum_{k=1}^{k=r} a_k S(n-k) \tag{2.63}$$

As the amplitude of the input signal is negligible, we can approximate $S(n)$ as $S(n) \cong \sum_{k=1}^{k=r} a_k S(n-k)$. This equation is known as prediction equation because nth sample of the speech signal is predicted using the past r samples of the identical speech signal. The co-efficiets $a_k \forall k = 1 \cdots n$ are known as Linear Predictive Co-efficients (LPC). The LPC completely describes the vocal tract filter. These are obtained from the speech signal using the following techniques.

2.4.1 Autocorrelation Method

The LPC's are obtained such that

$$E((S(n) - \sum_{k=1}^{k=r} a_k S(n-k))^2) \tag{2.64}$$

is minimized. In this E is the expectation operator and $(S(n) - \sum_{k=1}^{k=r} a_k S(n-k))^2$ is the squared error obtained in predicting the nth sample of the speech signal using the past r samples. minimizing (2.64) is achieved by partial differentiating the (2.64) with respect to unknown variables $a_j \forall j = 1 \cdots r$ and equate to zeros as mentioned below

$$\frac{\partial E((S(n) - \sum_{k=1}^{k=r} a_k S(n-k))^2)}{\partial a_j} \tag{2.65}$$

$$\Rightarrow E(2(S(n) - \sum_{k=1}^{k=r} a_k S(n-k))S(n-j)) = 0 \tag{2.66}$$

$$\Rightarrow R_S(j) = \sum_{k=1}^{k=r} a_k R_S(j-k) \tag{2.67}$$

$R_S(j)$ is the autocorrelation of the speech signal. The speech signal under consideration for the particular duration is assumed to be Wide Sense Stationary (W.S.S) and hence the autocorrelation depends on the difference of the index. As speech signal is the real signal and W.S.S, autocorrelation is symmetric function. This technique is known as autocorrelation method.

2.4.1.1 Solving (2.67) Using Levinson–Durbin Algorithm

The auto correlation $R_S(j)$ is computed as $E(S(n)S(n-j))$. Consider $S(n)$ is the random variable obtained by sampling across the random process S at the time instant n and $S(n-j)$ is the random variable obtained by sampling across the random process S at the time instant $n-j$. To compute $E(S(n)S(n-j))$, we need the joint probability density function $f_{S(n)S(n-j)}(\alpha, \beta)$. This is not available in practice. Hence the $R_S(j)$ is estimated from the sample speech signal itself. The estimation is done along the process assuming that the speech signal is ergodic in autocorrelation as follows

$$R_S(j) = \sum_{n=-\infty}^{n=\infty} S(n)S(n-j) \tag{2.68}$$

In practice, the computation is done for the longer duration of above 20 ms. The (2.68) is written in the matrix form for $r = 5$ as follows

$$\begin{bmatrix} R_S(1) \\ R_S(2) \\ R_S(3) \\ R_S(4) \\ R_S(5) \end{bmatrix} = \begin{bmatrix} R_S(0) & R_S(-1) & R_S(-2) & R_S(-3) & R_S(-4) \\ R_S(1) & R_S(0) & R_S(-1) & R_S(-2) & R_S(-3) \\ R_S(2) & R_S(1) & R_S(0) & R_S(-1) & R_S(-2) \\ R_S(3) & R_S(2) & R_S(1) & R_S(0) & R_S(-1) \\ R_S(4) & R_S(3) & R_S(2) & R_S(1) & R_S(0) \end{bmatrix} \begin{bmatrix} a_1 \\ a_2 \\ a_3 \\ a_4 \\ a_5 \end{bmatrix}$$

Due to symmetric nature of autocorrelation function, the equation in the matrix form is rewritten with $R_s(-n) = R_s(n)$.

Fig. 2.4 Matrix highlighting the toeplitz structure

$$\begin{bmatrix} R_S(0) & R_S(1) & R_S(2) & R_S(3) & R_S(4) \\ R_S(1) & R_S(0) & R_S(1) & R_S(2) & R_S(3) \\ R_S(2) & R_S(1) & R_S(0) & R_S(1) & R_S(2) \\ R_S(3) & R_S(2) & R_S(1) & R_S(0) & R_S(1) \\ R_S(4) & R_S(3) & R_S(2) & R_S(1) & R_S(0) \end{bmatrix}$$

The autocorrelation matrix thus obtained is the toeplitz matrix because, it is the symmetric matrix with the identical diagonal elements. (refer Fig. 2.4).

2.4.1.2 Levinson–Durbin Algorithm

Consider the vocal tract with 5 Linear predictive co-efficients (5th order LPC) (represented as $\overline{x_5^4} = [x_5(0)\ x_5(1)\ x_5(2)\ x_5(3)\ x_5(4)]^T$) are obtained by solving the equation mentioned in Fig. 2.5. If the vocal tract is modelled with 4 LPC (4th order LPC), the co-efficients $\overline{x_4^3} = [x_4(0)\ x_4(1)\ x_4(2)\ x_4(3)]$ are obtained by solving the equation mentioned in Fig. 2.6. The key idea in Levinson–Durbin algorithm is to obtain the 5th order LPC from the 4th order LPC. They are related as follows. Let $[c_5(0)c_5(1)c_5(2)c_5(3)c_5(4)]^T$ is the correction vector. Note that $x_5(4) = c_4(4)$.

$$\begin{bmatrix} x_5(0) \\ x_5(1) \\ x_5(2) \\ x_5(3) \\ x_5(4) \end{bmatrix} = \begin{bmatrix} x_4(0) \\ x_4(1) \\ x_4(2) \\ x_4(3) \\ 0 \end{bmatrix} + \begin{bmatrix} c_4(0) \\ c_4(1) \\ c_4(2) \\ c_4(3) \\ c_4(4) \end{bmatrix} \tag{2.69}$$

Representing the equation in Fig. 2.5 using (2.69), we get the following

Fig. 2.5 Equation for obtaining 5 LPC

$$\left[\begin{array}{cccc|c} a_0 & a_1 & a_2 & a_3 & a_4 \\ a_1 & a_0 & a_1 & a_2 & a_3 \\ a_2 & a_1 & a_0 & a_1 & a_2 \\ a_3 & a_2 & a_1 & a_0 & a_1 \\ \hline a_4 & a_3 & a_2 & a_1 & a_0 \end{array}\right] \left[\begin{array}{c} x_5(0) \\ x_5(1) \\ x_5(2) \\ x_5(3) \\ \hline x_5(4) \end{array}\right] = \left[\begin{array}{c} a_1 \\ a_2 \\ a_3 \\ a_4 \\ \hline a_5 \end{array}\right]$$

$$\begin{bmatrix} A_3 & \overline{a_4^r} \\ \overline{a_4^r}^T & a_0 \end{bmatrix} \begin{bmatrix} \overline{x_5^3} \\ x_5(4) \end{bmatrix} = \begin{bmatrix} \overline{a_4} \\ a_5 \end{bmatrix}$$

Fig. 2.6 Equation for obtaining 4 LPC

$$\left[\begin{array}{ccc|c} a_0 & a_1 & a_2 & a_3 \\ a_1 & a_0 & a_1 & a_2 \\ a_2 & a_1 & a_0 & a_1 \\ \hline a_3 & a_2 & a_1 & a_0 \end{array}\right] \left[\begin{array}{c} x_4(0) \\ x_4(1) \\ x_4(2) \\ \hline x_4(3) \end{array}\right] = \left[\begin{array}{c} a_1 \\ a_2 \\ a_3 \\ \hline a_4 \end{array}\right]$$

$$\begin{bmatrix} A_2 & \overline{a_3^r} \\ \overline{a_3^r}^T & a_0 \end{bmatrix} \begin{bmatrix} \overline{x_4^2} \\ x_4(3) \end{bmatrix} = \begin{bmatrix} \overline{a_3} \\ a_4 \end{bmatrix}$$

$$\begin{bmatrix} a_0 & a_1 & a_2 & a_3 & a_4 \\ a_1 & a_0 & a_1 & a_2 & a_3 \\ a_2 & a_1 & a_0 & a_1 & a_2 \\ a_3 & a_2 & a_1 & a_0 & a_1 \\ a_4 & a_3 & a_2 & a_1 & a_0 \end{bmatrix} \begin{bmatrix} x_4(0) + c_4(0) \\ x_4(1) + c_4(1) \\ x_4(2) + c_4(2) \\ x_4(3) + c_4(3) \\ 0 + c_4(4) \end{bmatrix} = \begin{bmatrix} a_1 \\ a_2 \\ a_3 \\ a_4 \\ a_5 \end{bmatrix} \tag{2.70}$$

Using the notations used in Fig. 2.5, (2.70) is represented as the following. Also let $c_4(4) = k_4$.

$$A_3\overline{x_4^3} + A_3\overline{c_4^3} + \overline{a_4^r}c_4(4) = \overline{a_4} \tag{2.71}$$

It is noted $A_3\overline{x_4^3} = \overline{a_4}$ and hence

$$A_3\overline{c_4^3} = -\overline{a_4^r}k_4 \tag{2.72}$$

$$\Rightarrow \overline{c_4^3} = -A_3^{-1}\overline{a_4^r}k_4 \tag{2.73}$$

It is also noted the following from the Fig. 2.5.

$$\overline{a_4^r}^T\overline{x_4^3} + \overline{a_4^r}^T\overline{c_4^3} + a_0k_4 = a_5 \tag{2.74}$$

$$\Rightarrow \overline{a_4^r}^T\overline{x_4^3} + \overline{x_4^{3r}}^T A_3^T\overline{c_4^3} + a_0k_4 = a_5 \tag{2.75}$$

$$\Rightarrow \overline{a_4^r}^T\overline{x_4^3} + \overline{x_4^{3r}}^T A_3c_4^3 + a_0k_4 = a_5 \tag{2.76}$$

Table 2.1 List of notations

Symbol	Notations
A	Area of cross section (m^2)
P	Pressure (Kg/ms^2)
S	Volume flow (m^3/s)
V	Velocity of the sound wave (m/s)
L	Volume (m^3)
ρ	Density of the air
m	Mass of the air
M	Molecular mass of the air
R	Gas constant
T	Temperature
c	Speed of the air
γ	Adiabatic constant
x	Distance from the origin on the x-axis (m)
t	Time (s)

Using (2.73), we get the following

$$\Rightarrow \overline{a_4^r}^T \overline{x_4^3} - \overline{x_4^{3r}}^T \overline{a_4^r} k_4 + a_0 k_4 = a_5 \tag{2.77}$$

$$k_4 = \frac{a_5 - \overline{a_4^r}^T \overline{x_4^3}}{a_0 - \overline{x_4^{3r}}^T \overline{a_4^r}} \tag{2.78}$$

Thus the Levinson–Durbin algorithm is described by (2.70), (2.73) and (2.78) and steps involved are summarized as follows

1. $a_0 x_1(0) = a_1, \Rightarrow x_1(0) = \overline{x_1^0} = \frac{a_1}{a_0}$
2. $\begin{bmatrix} a_0 & a_1 \\ a_1 & a_0 \end{bmatrix} \begin{bmatrix} x_2(0) \\ x_2(1) \end{bmatrix} = \begin{bmatrix} a_1 \\ a_2 \end{bmatrix}$
   $\begin{bmatrix} x_2(0) \\ x_2(1) \end{bmatrix} = \begin{bmatrix} x_1(0) \\ 0 \end{bmatrix} + \begin{bmatrix} c_1(0) \\ c_1(1) \end{bmatrix}$
3. Compute $k_1 = c_1(1) = \frac{a_2 - \overline{a_1^r}^T \overline{x_1^0}}{a_0 - \overline{x_1^{0r}}^T \overline{a_1^r}^T} = \frac{a_2 - a_1 x_1(0)}{a_0 - x_1(0) a_1}$
4. Compute $\overline{c_1^0} = -A_0^{-1} \overline{a_1^r} k_1 = -a_0^{-1} a_1 k_1$
5. Compute $\overline{x_2^1} = \begin{bmatrix} x_2(0) \\ x_2(1) \end{bmatrix} = \begin{bmatrix} x_1(0) \\ 0 \end{bmatrix} + \begin{bmatrix} c_1(0) \\ c_1(1) \end{bmatrix}$
6. $\begin{bmatrix} a_0 & a_1 & a_2 \\ a_1 & a_0 & a_1 \\ a_2 & a_1 & a_0 \end{bmatrix} \begin{bmatrix} x_2(0) \\ x_2(1) \\ x_2(2) \end{bmatrix} = \begin{bmatrix} a_1 \\ a_2 \\ a_3 \end{bmatrix}$
   $\begin{bmatrix} x_3(0) \\ x_3(1) \\ x_3(2) \end{bmatrix} = \begin{bmatrix} x_2(0) \\ x_2(1) \\ 0 \end{bmatrix} + \begin{bmatrix} c_2(0) \\ c_2(1) \\ c_2(2) \end{bmatrix}$
7. Compute $k_2 = c_2(2) = \frac{a_3 - \overline{a_2^r}^T \overline{x_2^1}}{a_0 - \overline{x_2^{1r}}^T \overline{a_2^r}^T}$
8. Compute $\overline{c_2^1} = -A_1^{-1} \overline{a_2^r} k_2$

9 Compute $\overline{x_3^2}$

10 In general, compute $k_i = c_i(i) = \frac{a_{i+1} - \overline{a_i^r}^T \overline{x_i^{i-1}}}{a_0 - \overline{x_i^{(i-1)r}}^T \overline{a_i^r}^T}$

11. Compute, $\overline{c_i^{i-1}} = -A_i^{-1} \overline{a_i^r} k_i$
12. Compute $\overline{x_i^{i-1}}$
13. Repeat the steps 10, 11, 12 for $i = 1 \cdots n$ to obtain the nth order LPC $\overline{x_n^{n-1}}$.

2.4.2 Auto Covariance Method

The LPC obtained using the Autocorrelation method needs long duration of speech signal(> 20ms) and hence the vocal tract model is not very accurate. But the computation time to obtain the LPC has been reduced by using Levinson–Durbin algorithm. More accurate vocal tract model is obtained for every 2ms speech signal data. This is obtained using covariance method as described below. The (2.66) is rewritten again for clarity using expectation operator as follows

$$E(2(S(n) - \sum_{k=1}^{k=r} a_k S(n-k))S(n-j)) = 0 \tag{2.79}$$

$$\Rightarrow E(S(n)S(n-j)) = \sum_{k=1}^{k=r} a_k E(S(n-k)S(n-j)) \tag{2.80}$$

In autocorrelation method, $E(S(n-k)S(n-j))$ is computed as $\sum_{n=-\infty}^{n=\infty} S(n-k)S(n-j)$ (In practice for the long duration of greater than 20 ms) and hence can be represented as $R_S(j-k)$. But in case of auto covariance method, $E(S(n-k)S(n-j))$ is computed as $\sum_{n=0}^{n=L-1} S(n-k)S(n-j) = C_{kj} = \sum_{n=0}^{n=L-1} S(n-j)S(n-k) = C_{jk}$. Hence the LPC using the co-variance method is computed by solving (2.81).

$$C_{0j} = \sum_{k=1}^{k=r} a_k C_{kj} \tag{2.81}$$

The (2.81) for $r = 5$ is represented as follows

$$\begin{bmatrix} C_{01} \\ C_{02} \\ C_{03} \\ C_{04} \\ C_{05} \end{bmatrix} = \begin{bmatrix} C_{11} & C_{12} & C_{13} & C_{14} & C_{15} \\ C_{12} & C_{22} & C_{23} & C_{24} & C_{25} \\ C_{13} & C_{23} & C_{33} & C_{34} & C_{35} \\ C_{14} & C_{24} & C_{34} & C_{44} & C_{45} \\ C_{15} & C_{25} & C_{35} & C_{45} & C_{55} \end{bmatrix} \begin{bmatrix} a_1 \\ a_2 \\ a_3 \\ a_4 \\ a_5 \end{bmatrix} \tag{2.82}$$

Note that the matrix in (2.82) is the symmetric matrix, but not the toeplitz matrix. Hence Levison–Durbin cannot be used to solve the (2.81). It is also noted that the diagonal elements are greater than the other elements of the matrix and hence the matrix is positive-semi-definite matrix. Hence this can be solved using diagonalization of the matrix or Gauss-elimination method. Note that positive-semi definite symmetric matrix is always diagonalizable (refer Appendix C) with non-negative eigenvalues as the diagonal elements of the diagonal matrix. The computation time required to solve (2.81) is greater than the time required to solve (2.67).

2.5 Lattice Structure to Obtain Excitation Source for the Typical Speech Signal

The transfer function of the vocal tract is modelled as Nth order all pole filter which is represented as $V(Z)$.

$$V(Z) = \frac{1}{1 + \sum_{k=1}^{k=N} a_k z^{-k}} \tag{2.83}$$

If the $E(Z)$ is the excitation source signal and $S(Z)$ is the output signal, they are related as $S(Z) = E(Z)V(Z)$. In discrete domain they are related as $e(n) = s(n) + \sum_{k=1}^{k=N-1} a_k s(n-k)$. The excitation source $e(n)$ needs the FIR filter coefficients a_k. This can also be realized using lattice structure. The excitation source computed for mth order filter is obtained directly from the excitation source for $(m-1)$th filter. Hence fixing up the order of the model becomes easier in real time in modelling the vocal tract filter. The lattice structure for the $1st$ order filter is as given in Fig. 2.8. Let $f_0(n) = g_0(n) = s(n)$ and $f_1(n) = e^1(n)$. Note that $e^i(n)$ is the nth sample of the excitation source with ith order model. They are related as follows

$$f_1(n) = f_0(n) + k_1 g_0(n-1), g_1(n) = k_1 f_0(n) + g_0(n-1) \tag{2.84}$$

The relationship using the first order filter is given as follows

$$e^1(n) = s(n) + a_1^1 s(n-1) \tag{2.85}$$

Comparing (2.84) and (2.85), we get $k_1 = a_1^1$. For the second order filter, we get the following

$$f_2(n) = f_1(n) + k_2 g_1(n-1) \tag{2.86}$$

$$\Rightarrow f_2(n) = f_0(n) + k_1 g_0(n-1) + k_2(g_0(n-2) + k_1 f_0(n-1)) \tag{2.87}$$

$$= s(n) + k_1 s(n-1) + k_2 s(n-2) + k_1 k_2 s(n-1) \tag{2.88}$$

$$g_2(n) = g_1(n-1) + k_2 f_1(n) \tag{2.89}$$

The relationship using the second order filter is given as follows

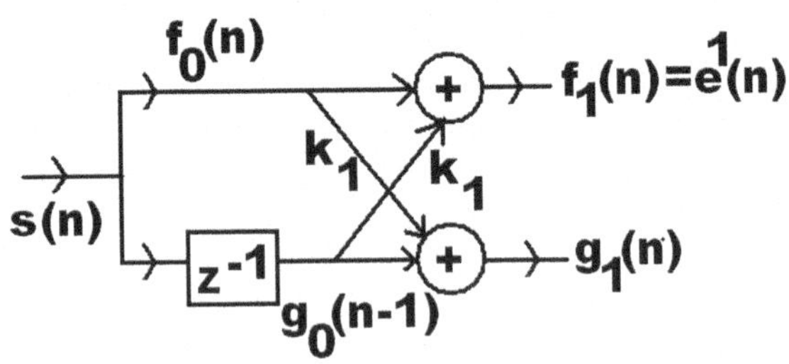

Fig. 2.7 Lattice Structure for the first order filter

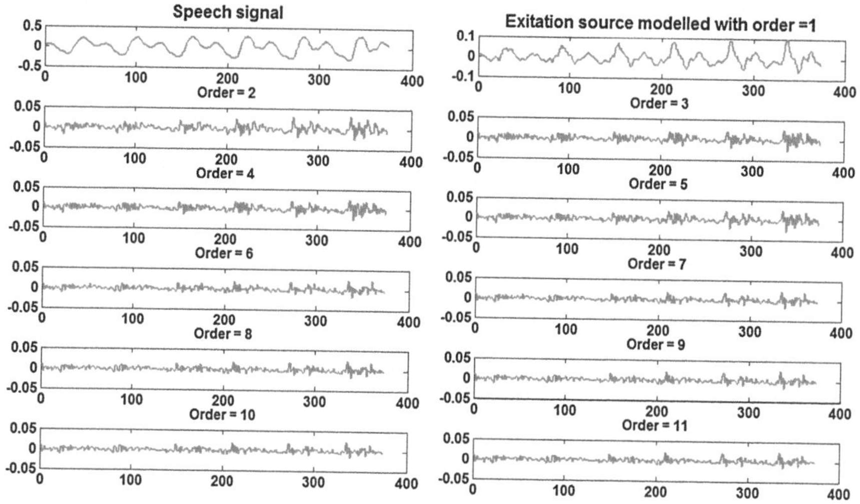

Fig. 2.8 Speech signal and the corresponding excitation source modelled using lpc with different co-efficients

$$f_2(n) = e^2(n) = s(n) + a_1^2 s(n-1) + a_2^2 s(n-2) \tag{2.90}$$

Comparing (2.88) and (2.90), we get the following

$$a_1^2 = k_1 + k_1 k_2 \tag{2.91}$$
$$a_2^2 = k_2 \tag{2.92}$$

In general it is noted that $a_r^r = k_r$. Also it is noted that $g_2(n) = k_1 f_0(n-1) + g_0(n-2) + k_2(f_0(n) + k_1 g_0(n-1))$, which is simplified as follows

$$g_2(n) = k_2 s(n) + (k_1 + k_1 k_2)s(n-1) + s(n-2) \tag{2.93}$$
$$g_2(n) = a_2^2 s(n) + a_1^2 s(n-1) + s(n-2) \tag{2.94}$$

Comparing (2.90) and (2.94), we understand that if the filter co-efficients of $f_2(n)$ are arranged in the reverse order, we get the filter co-efficients for $g_2(n)$. In z-domain $G_2(Z) = Z^{-2} F_2(Z^{-1})$. In general

$$G_N(Z) = Z^{-N} F_N(Z^{-1}) \tag{2.95}$$

2.5.1 Computation of Lattice Co-efficient from LPC Co-efficients

Let the Nth order transfer function $\frac{1}{V(Z)}$ is represented as $A_N(Z)$.

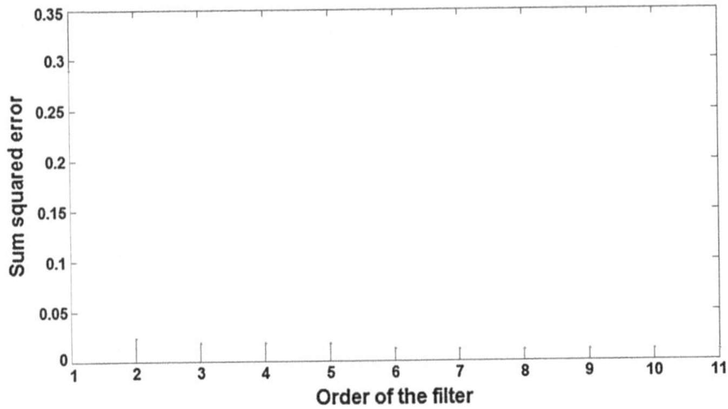

Fig. 2.9 Sum squared error (sum squared value of the samples of the excitation source) obtained using lpc model versus order of the lpc filter (number of lattice co-efficients)

$$A_N(Z) = 1 + a_1^N z^{-1} + a_2^N z^{-2} + a_3^N z^{-3} + \cdots + a_N^N z^{-N} \tag{2.96}$$

Note that a_N^N is the lattice co-efficient k_N. So if the $(N-1)$th order transfer function $A_{N-1}(Z)$ is obtained , the co-efficient of z^{N-1} of $A_{N-1}(Z)$ is obtained as k_N. The relation between $A_N(Z)$ and $A_{N-1}(Z)$ is needed and are obtained as follows. The excitation source signals obtained for various order and the corresponding sum squared values are displayed in Figs. 2.8 and 2.9 respectively.

$$F_N(Z) = F_{N-1}(Z) + k_N G_{N-1}(Z) Z^{-1} \tag{2.97}$$

$$G_N(Z) = k_N F_{n-1}(Z) + G_{N-1}(Z) Z^{-1} \tag{2.98}$$

$$S(Z) A_N(Z) = E_N(Z) = F_N(Z) \tag{2.99}$$

$$\Rightarrow A_N(Z) = A_{N-1}(Z) + k_N \frac{G_{N-1}(Z)}{S(Z)} Z^{-1} \tag{2.100}$$

Let $B_N(Z) = \frac{G_N(Z)}{S(Z)}$ and we get the following.

$$A_N(Z) = A_{N-1}(Z) + k_N B_{N-1}(Z)^{-1} \tag{2.101}$$

$$B_N(Z) = k_N A_{N-1}(Z) + B_{N-1}(Z) Z^{-1} \tag{2.102}$$

$$\Rightarrow A_N(Z) = A_{N-1}(Z) + k_N (B_N(Z) - k_N A_{N-1}(Z)) \tag{2.103}$$

$$\Rightarrow A_{N-1}(Z) = \frac{A_N(Z) - k_N B_N(Z)}{1 - k_N^2} \tag{2.104}$$

Note that $B_0(Z) = \frac{G_0(Z)}{S(Z)} = 1$. Thus the steps involved in computing lattice parameters from Nth order lpc are summarized as follows.

```
function [L,k,E,ERROR]=lattice(S1,FS)
L=lpc(S1,11);
k=[];
for i=1:1:length(L)-1
    k=[k L(length(L))];
    R=L(length(L):-1:1);
temp=(L-k(i)*R)/(1-(k(i)^2))
temp1=temp(1:1:length(temp)-1);
L=temp1;
end
close all
k=k(length(k):-1:1);
ERROR=[];
r=1;
for j=4:1:12
e=[];
for i=1:1:11
    f{i}(1)=0;
    g{i}(1)=0;
end
for n=2:1:length(S1)
f{1}(n)=S1(n);
g{1}(n)=S1(n);
g{1}(n-1)=S1(n-1);
for i=2:1:j
f{i}(n)=f{i-1}(n)+k(i-1)*g{i-1}(n-1) ;
g{i}(n)=g{i-1}(n-1)+k(i-1)*f{i-1}(n);
end
e=[e f{j}(n)];
end
E{r}=e;
ERROR=[ERROR sum(E{r}.^2)];
r=r+1;
end
figure
stem([2:1:11],ERROR)
figure
for i=2:1:10
    subplot(5,2,i)
    plot(E{i})
end
subplot(5,2,1)
plot(S1)
```

1. Obtain $A_N(Z) = 1 + a_1^N z^{-1} + a_2^N z^{-2} + a_3^N z^{-3} + \cdots + a_N^N z^{-N}$ using the lpc
2. $K_N = a_N^N$
3. Compute $B_N(Z) = Z^{-N} A_N(Z^{-1})$. Trick is to arrange the co-efficients of $A_N(Z)$ in the reverse order to obtain $B_N(Z)$.
4. Compute $A_{N-1}(Z) = \frac{A_N(Z) - k_N B_N(Z)}{1 - k_N^2}$. Identify the co-efficient of $(N-1)$ to obtain K_{N-1}
5. Repeat 3 and 4 to obtain the lattice co-efficients.

```
%levinsondurbin.m
function [res]=levinsondurbin(a)
%a is the vector with size 1xn
temp1=a(2:1:length(a));
for j=1:1:length(a)-1
A{j}=toeplitz(a(1:1:j));
end
x{1}=[a(2)/a(1)];
k(1)=(a(3)-a(2)*x{1}(1))/(a(1)-a(2)*x{1}(1));
c{1}=-inv(A{1})*a(2)*k(1) ;
c{1}=[c{1};k(1)];
x{2}=[x{1} ;0];
x{2}=x{2}+c{1};
for r=2:1:length(a)-2

k(r)=(a(r+2)-rev(a,r)*x{r})/(a(1)-[a(2:1:r+1)]*x{r});
c{r}=-1*inv(A{r})*rev(a,r)' *k(r) ;
c{r}=[c{r};k(r)];
x{r+1}=[x{r};0];
x{r+1}=x{r+1}+c{r};
end
res=[1; -1*x{r+1}];
```

Chapter 3
Feature Extraction of the Speech Signal

Abstract Isolated speech recognition, speaker recognition, and continuous speech recognition require the feature vector extracted from the speech signal. This is subjected to pattern recognition to formulate the classifier. The feature vector is extracted from each frame of the speech signal under test. In this chapter, various parameter extraction techniques such as linear predictive co-efficients as the filter co-efficients of the vocal tract model, poles of the vocal tract filter, cepstrual co-efficients, mel-frequency cepstral co-efficients (MFCC), line spectral co-efficients, and reflection co-efficients are discussed in this chapter. The preprocessing techniques such as dynamic time warping, endpoint detection, and pre-emphasis are also discussed in this chapter.

3.1 Endpoint Detection

The isolated speech signal recorded through the microphone will have noise at both ends of the speech segment. There is the need to identify the beginning and the ending of the speech segment from the recorded speech signal. This is known as endpoint detection. This is identified as follows.

1. The speech signal S is divided into frames. Compute the sum-squared values of the individual frames. The energy of the frame consists of voiced speech signal (due to vibration of the vocal chords) is usually greater than the noise signal. Identify the first frame of the speech segment that has the energy greater than the predefined upper threshold value. From this point onwards, search the frame in the backward direction such that the energy of frame exceeds the predefined lower threshold value. Let the identified first frame of the voiced speech segment is represented as V.
2. Let $S(n)$ be the nth sample of the speech signal. If $sgn(S(n))sgn(S(n+1))$ is negative, zero crossing has happened at the position nth sample of the speech signal. The zero-crossing rate of the unvoiced segment near to the voiced segment

E. S. Gopi, *Digital Speech Processing Using Matlab*,
Signals and Communication Technology, DOI: 10.1007/978-81-322-1677-3_3,
© Springer India 2014

is larger when compared to the noise. The number of zero crossings per frame is known as zero-crossing rate. Once the first frame of the voiced speech segment (V) is identified using the energy computation, the first frame of the unvoiced speech segment (if available) available prior to V is identified as follows. From V, search the previous 25 frames backwards to choose the first frame that has the zero-crossing rate lesser than the predefined threshold value and it is declared as the first unvoiced speech frame.

3. The above procedure is repeated from the last sample of the speech segment to identify the endpoint of the speech segment.

```
%endpointdetection.m
function [res1,res2,speechsegment,utforste,
   ltforste,ltforzcr]...
=endpointdetection(S,FS)
%mzcr, mste-mean of the zero-crossing rate and the
short-time energy for the first 100\,ms
%vzcr,vste-variance of the zero-crossing rate and the
short-time energy for the first 100ms
%utforste-upper threshold for short-time energy
%ltforste-lower threshold for short-time energy
%ltforzcr-lower threshold for zero-crossing rate
%fl - frame length is fixed as 10 ms
fl=fix(FS/100);
%DC offset removal
S=S-mean(S);
tempdata=S(1:1:fl*10);
temp1=blkproc(tempdata,[1 fl],'zcr(x)');
mzcr=mean(temp1);
vzcr=var(temp1,1);
temp2=blkproc(tempdata,[1 fl],'ste(x)');
mste=mean(temp2);
vste=var(temp2,1);
ltforste=mste*100-(sqrt(vste)/10);
utforste=mste*100+(sqrt(vste)/10);
ltforzcr=mzcr*100-(sqrt(vzcr)/10);
res1=blkproc(S,[1 fl],'zcr(x)');
res2=blkproc(S,[1 fl],'ste(x)');
figure(1)
subplot(2,1,1)
 plot(res1)
 xlabel('frame number')
 ylabel('Zero-crossing rate')
 subplot(2,1,2)
 plot(res2)
 xlabel('frame number')
```

```
ylabel('Short-time energy')

[p1,q1]=find(res2>utforste);
temp3=res2(q1(1):-1:1);
[p2,q2]=find(temp3<ltforste);
if(isempty(q2)==1)
    q2(1)=0;
temp4=res1(q1(1)-q2(1):-1:1);
else
temp4=res1(q1(1)-q2(1):-1:1);
end
[p3,q3]=find(temp4<ltforzcr);
res2rev=res2(length(res2):-1:1);
[p4,q4]=find(res2rev>utforste);
temp5=res2rev(q4(1):-1:1);
[p5,q5]=find(temp5<ltforste);
res1rev=res1(length(res1):-1:1);
if(isempty(q5)==1)
    q5(1)=0;
temp6=res1rev(q4(1)-q5(1):-1:1);
else
temp6=res1rev(q4(1)-q5(1):-1:1);
end
[p6,q6]=find(temp6<ltforzcr);
speechsegment=S((length(temp4)-q3(1)+1)*fl:1:
length(S)...
-(length(temp6)-q6(1)+1)*fl);
figure
subplot(2,1,1)
plot(S)
title('Original speech signal')
subplot(2,1,2)
plot(speechsegment)
title('Speech segment after endpoint detection')

%ste.m
function [res]=ste(x)
res=sum(x.^2)/length(x);

function res=zcr(x)
res=length(find(diff(cumsum(sign(x)))==-1));
```

3.2 Dynamic Time Warping

The speech segments corresponding to the identical words (even when spelled by the identical person) will not have the same length. Hence, the length of the speech segment should be preprocessed to maintain the uniform length. This is done using the dynamic time warping . Let the two speech segments corresponding to the particular identical word are represented as the column vector X and with length N_1 and N_2 samples, respectively. Let $\overline{X}$ is the reference speech segment and the speech segment $\overline{Y}$ is warped such that its length is made equal to N samples as described below.

1. Formulate the distance matrix D with size $N_1 \times N_2$. The (i, j)th element of the matrix D is $(X_i - Y_j)^2$ where X_i is the ith element of the vector X and the Y_j is the jth element of the vector Y.
2. The idea is to identify the path from $(1, 1)$ to (N_1, N_2) such the cumulative distance is minimized. The path assumes only one-step movement either rowwise, columnwise, or diagonalwise in the forward direction as shown in Fig. 3.1. This is achieved as follows.
3. Formulate the cumulative distance matrix C with the following rule
   a. $C(1, 1) = D(1, 1)$
   b. $C(i, j) = D(i-1, j-1) + \min(C(i, j-1), C(i-1, j), C(i-1, j-1)) \forall i$ or $j \neq 1$
   c. $C(1, j) = C(1, j-1) + D(1, j)$ and $C(j, 1) = C(j-1, 1) + D(j, 1)$
4. Back trace the path from the position (N_1, N_2) to $(1, 1)$ such that the cumulative distance is minimum. The path starts with the position (N_1, N_2). The next position is obtained as $(i, j) = arg\ min(C(N_1 - 1, N_2), C(N_1, N_2 - 1), C(N_1 - 1, N_2 - 1))$
   The illustration of dynamic time warping after endpoint detection for the typical speech segment are illustrated in the (Figs. 3.2, 3.3, 3.4, 3.5 and 3.6).

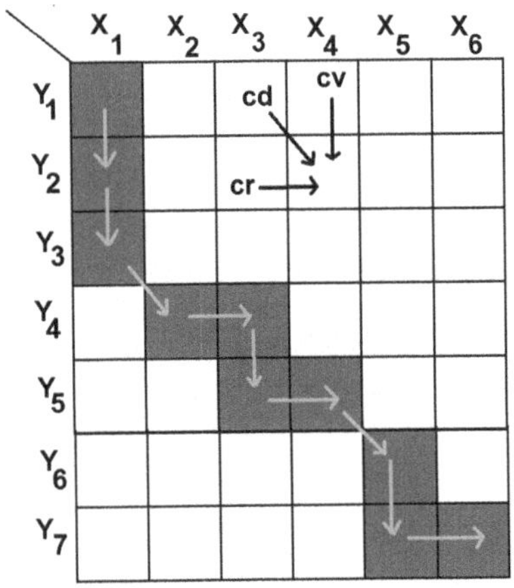

Fig. 3.1 Illustration of dynamic time warping with the typical shortest point obtained in the cumulative matrix. The cumulative shortest value at the position $(2, 4)$ is obtained as $min(cd, cv, cr) + D(2, 4)$ is also illustrated

```
function [temp]=dynamictimewarping(S1,S2)
%If S1 is the query speech
   signal and S2 is in the database
%after endpoint detection
figure
plot(S1)
hold on
plot(S2,'r')
title('Query and the reference speech signal after
   endpoint detection')
%Form the squared euclidean distance matrix
DC=zeros(size(S1,2),size(S2,2));
%Form the cumulative distance matrix
DC(1,1)=(S1(1)-S2(1))^2;
DC(1,2)=(S1(1)-S2(1))^2+(S1(1)-S2(2))^2;
DC(2,1)=(S1(1)-S2(1))^2+(S1(2)-S2(1))^2;
for i=3:1:length(S1)
    DC(i,1)=DC(i-1,1)+(S1(i)-S2(1))^2;
end
for i=3:1:length(S2)
    DC(1,i)=DC(1,i-1)+(S1(1)-S2(i))^2;
end
for i=2:1:length(S1)
```

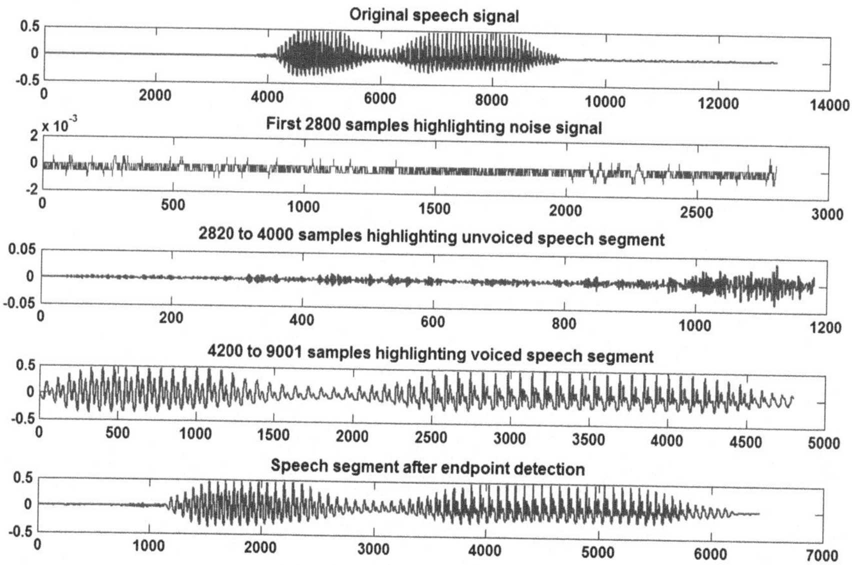

Fig. 3.2 Illustration of endpoint detection of the typical speech segment with the sampling frequency of 11250 Hz

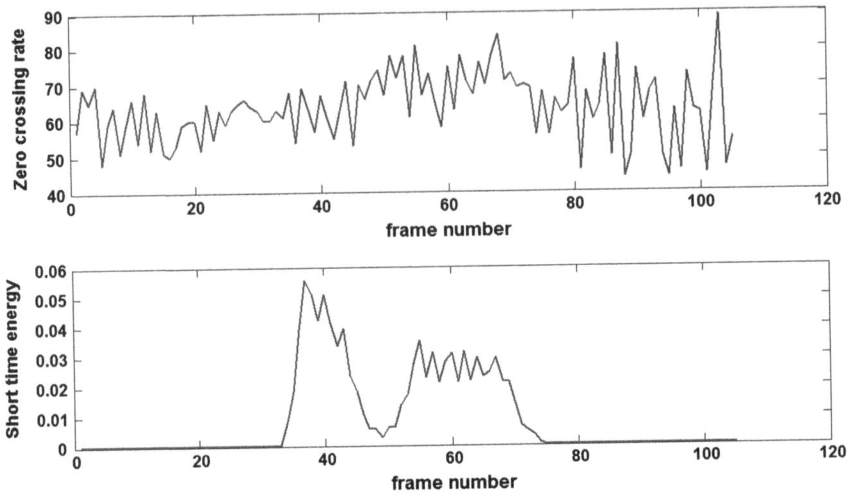

Fig. 3.3 Illustration of computations of short-term energy and the zero-crossing rate for endpoint detection of the typical speech segment mentioned in Fig. 1.1

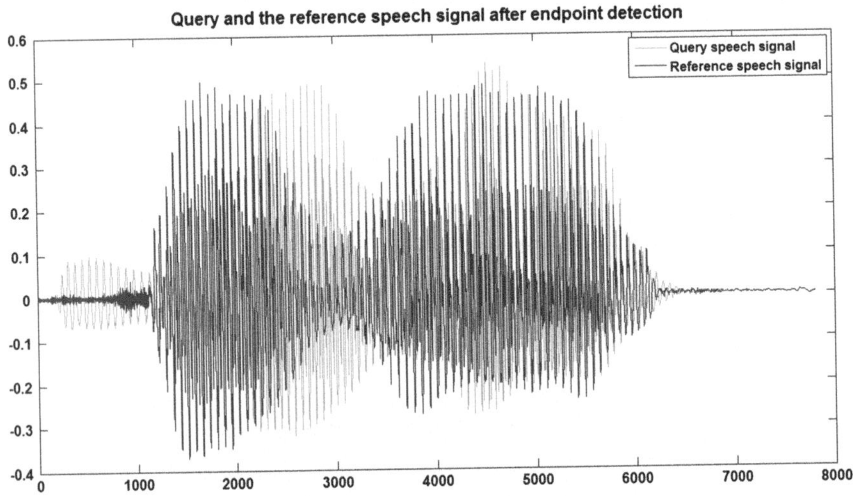

Fig. 3.4 Query speech signal (before DTW) and the reference speech signals after endpoint detection

```
    for j=2:1:length(S2)
        temp=[DC(i-1,j) DC(i,j-1) DC(i-1,j-1)];
        DC(i,j)=min(temp)+(S1(i)-S2(j))^2;
    end
end
%Argument collection
```

```
n=2;
POS{1}=[length(S1) length(S2)];
ITER=1;
i=length(S1);
j=length(S2);
while(ITER==1)
if((i==1)&(j~=1))
    POS{n}=[1 j-1];
    i=POS{n}(1);
    j=POS{n}(2);
    n=n+1;
    break
elseif((j==1)&(i~=1))
    POS{n}=[i-1 1];
    i=POS{n}(1);
    j=POS{n}(2);
    n=n+1;
    elseif((j==1)&(i==1))
    POS{n}=[1 1];
    ITER=2;
    n=n+1;
    break
else
    [p,q]=min([DC(i-1,j-1) DC(i-1,j) DC(i,j-1)])
    if(q==1)
     POS{n}=[i-1 j-1];
     elseif(q==2)
     POS{n}=[i-1 j];
     elseif(q==3)
     POS{n}=[i j-1];
    end
i=POS{n}(1);
j=POS{n}(2);
n=n+1;
end
end
%Speech warping
POS1=cell2mat(POS');
spos1=size(POS1,1);
POS1=[POS1;ones(1,POS1(spos1,2))'  (POS1(spos1,2):
     -1:1)'];
for i=1:1:length(S2)
    [p,q]=find(POS1(:,2)==i);
    [r,s]=min(abs(POS1(p,2)-i));
    temp(i)=S1(1,POS1(p(s)))
```

```
end
figure
subplot(3,1,1)
plot(S2)
title('Reference speech signal ')
subplot(3,1,2)
plot(S1,'r')
title('Query speech signal')
subplot(3,1,3)
plot(temp,'g')
title('Query speech signal after dynamic time warping')
```

3.3 Linear Predictive Co-efficients

Vocal tract is modeled as the all-pole filter (refer Chap. 2) and the transfer function of the rth-order all-pole filter is as given in (3.1).

$$V(Z) = \frac{GZ^{-\frac{r}{2}}}{1 - \sum_{k=1}^{k=r} a_k Z^{-k}} \tag{3.1}$$

$Z^{-\frac{r}{2}}$ contributes the delay introduced due to vocal tract and it does not contribute to the magnitude response. Hence, the transfer function of the vocal tract is represented as $\frac{G}{1-\sum_{k=1}^{k=r} a_k Z^{-k}}$ where G is the gain constant. The filter co-efficients a_k $\forall k = 1 \ldots r$ are described as linear predictive co-efficients. The LPC are obtained using auto-correlation method and the autocovariance method as described in the Chap. 2. Magnitude response of the vocal tract model obtained from every non-overlapping 30 ms speech segment (refer Fig. 3.6) is illustrated in Fig. 3.7. The LPC are obtained using autocorrelation method as described in the Chap. 2. Also the order of the filter and the gain G are chosen as 13 and 1, respectively (Fig. 3.8).

```
%vtlpc.m
function [lpcoef]=vtlpc(S,FS,r)
%lpc model for every 30\,ms speech segment after endpointdetection
%S is the speech signal, FS is the sampling frequency
%r is the order of the filter.
[res1,res2,S,utforste,ltforste,ltforzcr]=endpointdetection(S,FS);
nos=fix(length(S)/(FS*30*10^(-3)));
S=S(1:1:nos*FS*30*10^(-3));
[lpccoef]=blkproc(S,[1 FS*30*10^(-3)],'lpc1(x,P1)',r-1);

%lpc1.m
function [lpccoef]=lpc1(x,r)
%Autocorrelation  method for computing vocal tract linear
predictive co-efficients
for i=1:1:r
```

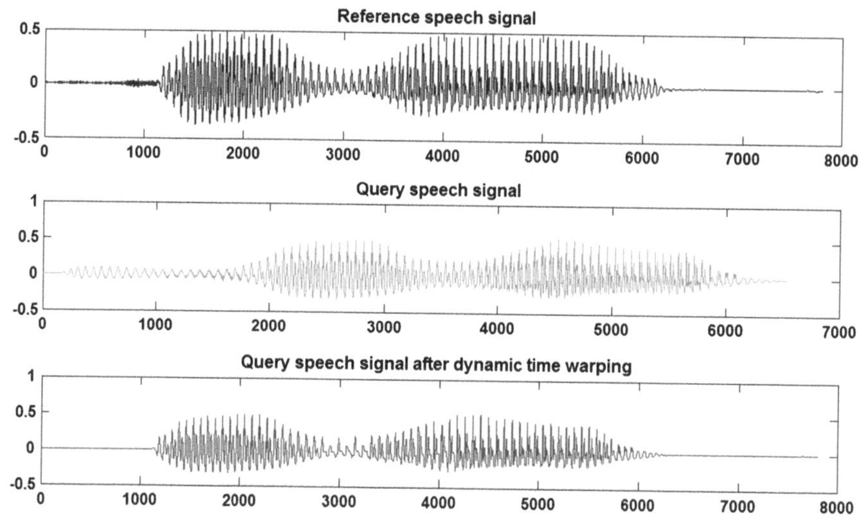

Fig. 3.5 Illustration of dynamic time warping. It is noted that the shape of the query speech signal looks like the original reference speech signal

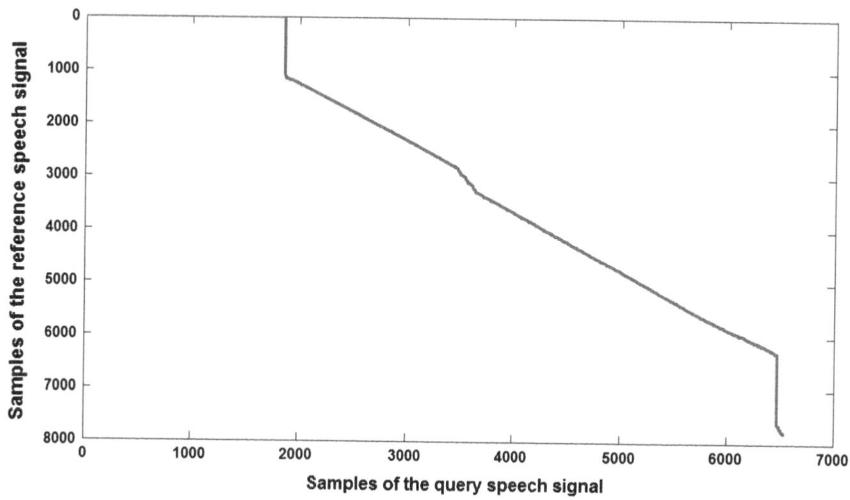

Fig. 3.6 Shortest path obtained using the dynamic time warping for the example mentioned in Fig. 3.5

```
    for j=1:1:r
    temp1=[zeros(1,i) x];
    temp2=[zeros(1,j) x];
m= max(size(temp1,2),size(temp2,2));
temp1=[temp1 zeros(1,m-length(temp1)+1)];
temp2=[temp2 zeros(1,m-length(temp2)+1)];
C(i,j)=sum(temp1.*temp2)
```

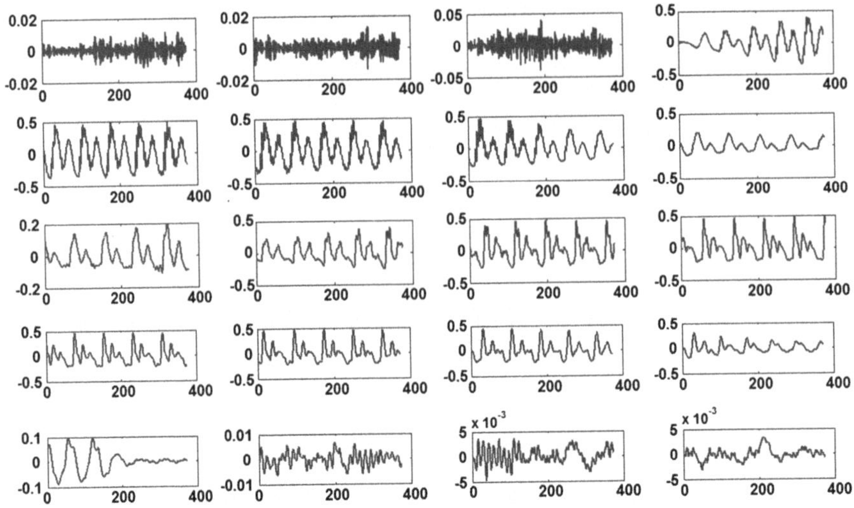

Fig. 3.7 Non-overlapping 30 ms speech segment with the sampling frequency of 12500 Hz

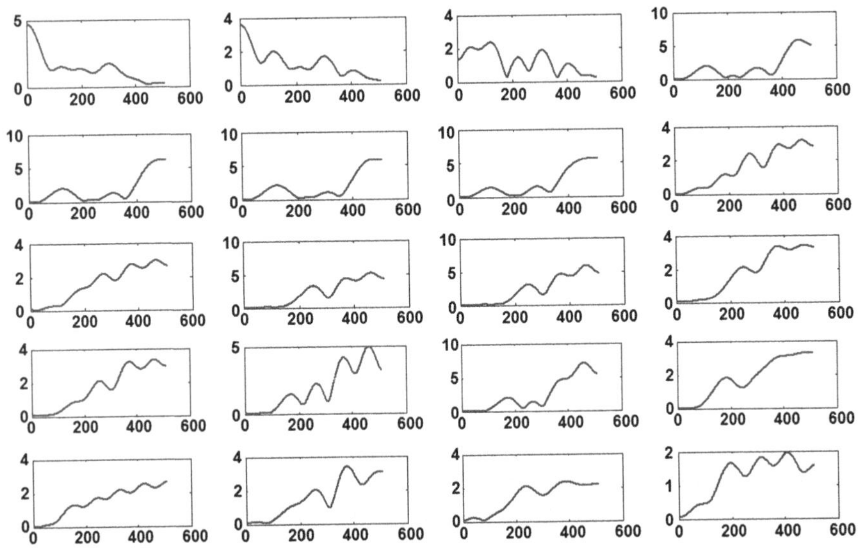

Fig. 3.8 Magnitude response of the vocal tract filter (13th order) obtained from the corresponding speech segment mentioned in Fig. 3.7

```
 end
end
for k=1:1:r
temp=[zeros(1,k) x];
x=[x zeros(1,k)];
C1(k)=sum(temp.*x);
```

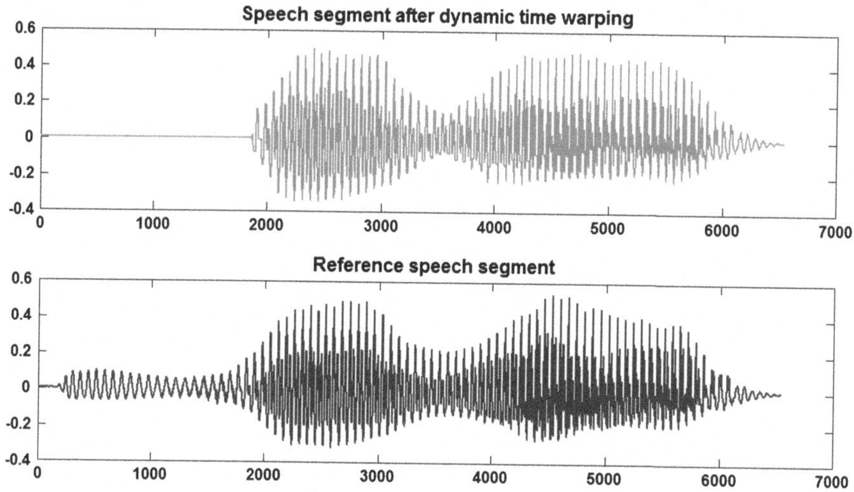

Fig. 3.9 Two speech segments corresponding to the identical words

```
end,
lpccoef=pinv(C)*C1';
lpccoef=[1;-1*lpccoef];
```

3.4 Poles of the Vocal Tract

The transfer function of the vocal tract is also represented as (3.2) where x_k are the roots of the polynomial $1 - \sum_{k=1}^{k=r} a_k Z^{-k}$. These are called poles of the vocal tract filter.

$$V(Z) = \frac{G}{\prod_{k=1}^{k=r}(1 - x_k Z^{-1})} \tag{3.2}$$

Two speech segments corresponding to the identical words are subjected to endpoint detection and dynamic time warping are as shown in Fig. 3.9. The LPC are collected for the individual frames from both the speech segments. They are displayed in Fig. 3.10. The roots of the corresponding vocal tract polynomial are displayed in Fig. 3.11. It is noted that the poles of the vocal tract model and the LPC obtained using the two speech segments are almost identical. It is noted that G is chosen as 1.

```
%rvt.m
function [res]=rvt(S1,S2,FS)
%roots of the vocal tract of the two speech segments
%corresponding to the identical word
[res1,res2,S11,utforste,ltforste,ltforzcr]=endpointdetection(S1,FS);
[res1,res2,S12,utforste,ltforste,ltforzcr]=endpointdetection(S2,FS);
S11=dynamictimewarping(S11,S12);
lpccoef11=mainprogramforlpc(S11,FS);
lpccoef12=mainprogramforlpc(S12,FS);
```

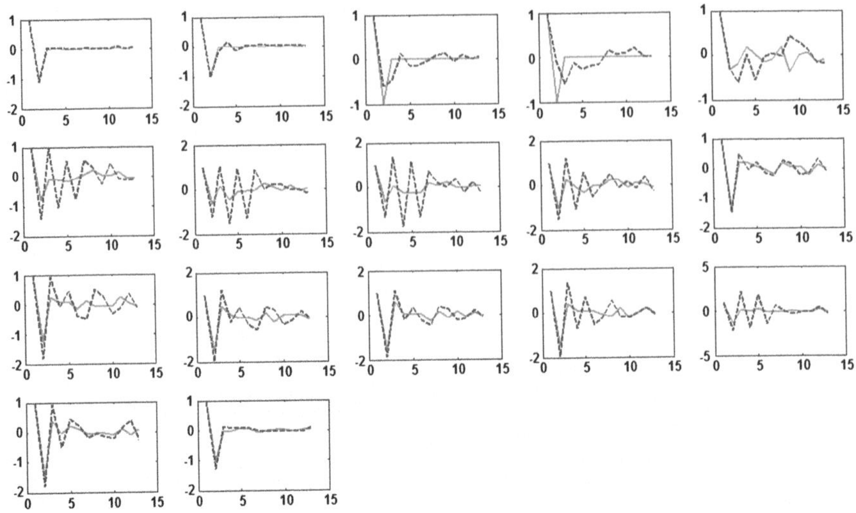

Fig. 3.10 LPC obtained for the individual frames of the two speech segments (corresponding to the identical words) mentioned in Fig. 3.9

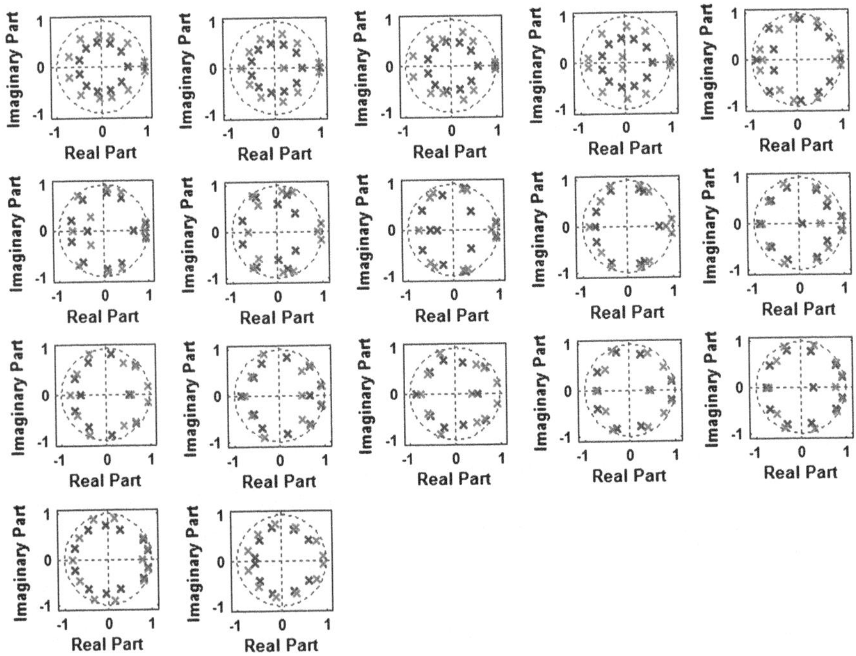

Fig. 3.11 Roots of the vocal tract filter (*poles*) of the individual frames of the two speech segments (corresponding to the identical words mentioned in Fig. 3.9)

```
for i=1:1:size(lpccoef11,2)
figure
subplot(4,5,i)
res=[roots(lpccoef11(:,i))  roots(lpccoef12(:,i))];
zplane([ ],res)
end

for i=1:1:size(lpccoef11,2)
figure
subplot(4,5,i)
plot(lpccoef11(:,i),'r')
hold on
plot(lpccoef12(:,i),'b--')
end
```

3.5 Reflection Co-efficients

The reflection co-efficients mentioned in (2.62) is related to the LPC in (2.63). From LPC, the reflection co-efficients are obtained. This can also be used as the speech features for further classification. The values are within the range -1 to $+1$. Interpolating the reflection co-efficients helps in getting the smooth frequency response. The negative reflection co-efficients are known as partial correlation (PARCOR).

3.6 Log Area Ratio

Negative log of (2.54) are computed for all valued of i and are treated as the attributes for speech feature vector. This is known as log area ratio (LAR) as mentioned below.

$$LAR = -\log\left(\frac{A_{i-1}}{A_i}\right) = \log(\frac{(1+r_i)}{(1-r_i)} \Rightarrow r_i = \frac{e^{LAR}-1}{e^{LAR}+1} \tag{3.3}$$

The main advantage of using this is that the stability is guaranteed. (i.e., the value of reflection coefficient is bounded to 1 for any value of LAR as shown in Fig.3.12.

3.7 Cepstrum

The $C(Z) = \log(V(Z))$ is known as the cepstrum and the corresponding discrete samples in the time domain is known as cepstral co-efficients. The cepstral co-efficients are computed as follows.

$$c_n = \frac{1}{2\pi}\int_0^{2\pi} \log(V(Z))dZ \tag{3.4}$$

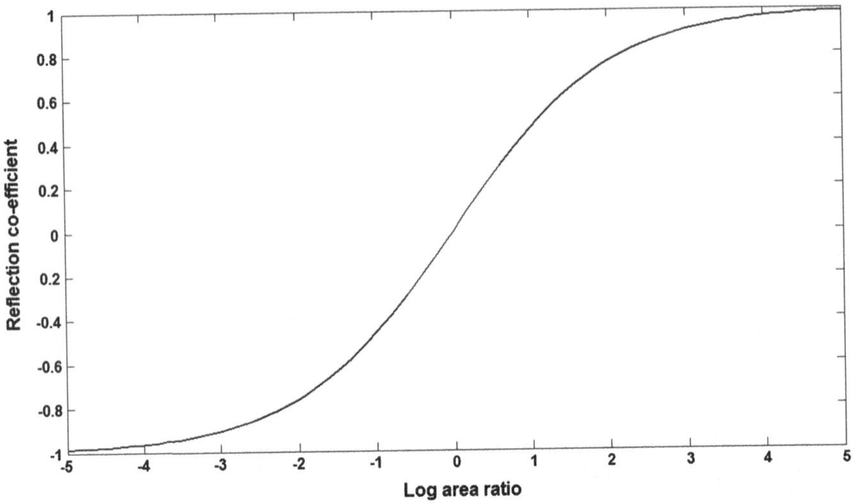

Fig. 3.12 Relationship between the reflection coefficient and the LAR

Substituting (3.2) in (3.4) with $G = 1$ we get the following.

$$c_n = \frac{1}{2\pi}\int_0^{2\pi} \log\left(\frac{1}{\prod_{k=1}^{k=r}(1 - x_k Z^{-1})}\right) \mathrm{d}Z \tag{3.5}$$

$$\Rightarrow c_n = -\frac{1}{2\pi}\int_0^{2\pi} \log\left(\prod_{k=1}^{k=r}(1 - x_k Z^{-1})\right) \mathrm{d}Z \tag{3.6}$$

Comparing $c_n = \frac{1}{2\pi}\int_0^{2\pi} C(Z)\mathrm{d}Z$ and (3.6), we get the following $C(Z) = -\log(\prod_{k=1}^{k=r}(1 - x_k Z^{-1})) = -\sum_{k=1}^{k=r}\log((1 - x_k Z^{-1}))$. If all the roots are within the unit circle (for stable system), $|x_k Z^{-1}| < 1$, and hence,

$$C(Z) = -\sum_{k=1}^{k=r}\log((1 - x_k Z^{-1})) = -\sum_{k=1}^{k=r}\sum_{n=0}^{n=\infty}\frac{(x_k Z^{-1})^n}{n} \tag{3.7}$$

$$\Rightarrow C(Z) = -\sum_{n=0}^{n=\infty}\sum_{k=1}^{k=r}\frac{(x_k Z^{-1})^n}{n} \tag{3.8}$$

Comparing $C(Z) = \sum_{n=-\infty}^{n=\infty} c_n Z^{-n}$ with (3.8), we get the cepstral co-efficients in terms of roots of the vocal tract filter (poles) as follows.

$$c_n = 0 \; \forall \; n < 0, \; c_n = \log G \; for \; n = 0, \; c_n = \sum_{k=1}^{k=r}\frac{x_k^n}{n} \forall \; n > 0 \tag{3.9}$$

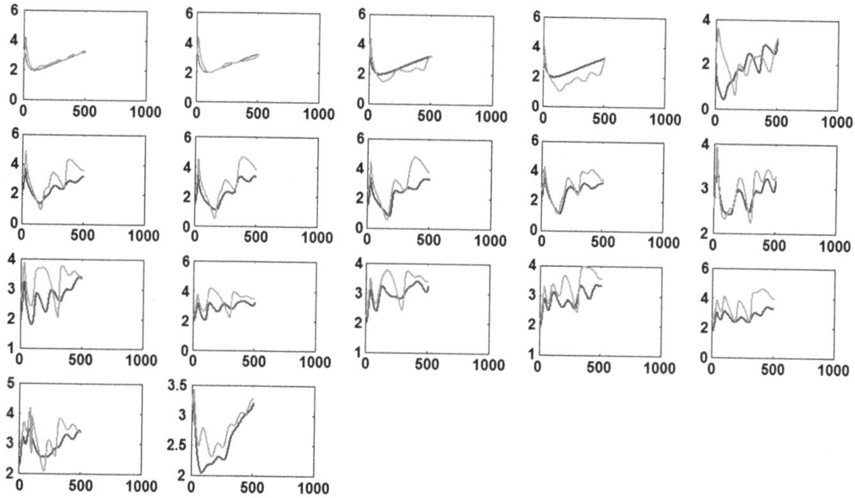

Fig. 3.13 Magnitude response of the cepstral co-efficients (cepstrum) of the individual frames of the two speech segment (corresponding to the identical words mentioned in Fig. 3.9)

It is noted that $\log_e(|r|e^{jw}) = \log|r| + jw$, and hence,

$$|C(Z)| = |\log V(Z)| = |\log|V(Z)| + j\angle(V(Z))| \tag{3.10}$$

The magnitude response of the cepstrum is computed for the two speech segments corresponding to the identical words (mentioned in Fig. 3.9) are mentioned in Fig. 3.13. This is computed using (3.10). The corresponding absolute values of the cepstral co-efficients computed using (3.9) is mentioned in Fig. 3.14. Note that the first cepstral coefficient is computed for $n = 0$ as $\log G = \log 1 = 0$, which is not mentioned in Fig. 3.14.

```
%cepstrumcoef.m
function [Cepstrum1,Cepstrum2,ccoef1,ccoef2]=cepstrumcoef(S1,S2,FS)
%Computation of magnitude response of the cepstrum and
%the corresponding cepstral co-efficients
[res1,res2,S11,utforste,ltforste,ltforzcr]=endpointdetection(S1,FS);
[res1,res2,S12,utforste,ltforste,ltforzcr]=endpointdetection(S2,FS);
S11=dynamictimewarping(S11,S12);
lpccoef11=mainprogramforlpc(S11,FS);
lpccoef12=mainprogramforlpc(S12,FS);
for i=1:1:size(lpccoef11,2)
figure(1)
subplot(4,5,i)
[H,W]=freqz(lpccoef11(:,i));
Cepstrum1{i}=abs(log(H)+j*W)
plot(Cepstrum{i})
subplot(4,5,i)
hold on
[H,W]=freqz(lpccoef12(:,i));
Cepstrum2{i}=abs(log(H)+j*W)
```

```
plot(Cepstrum{i},'r')
p=roots(lpccoef11(:,i));
for n=1:1:20
    temp1=0;
    for k=1:1:length(p)
    temp1=temp1+((p(k)^n)/n)
    end
    ccoef1{i}(n)=temp1/n;
end
q=roots(lpccoef12(:,i));
for n=1:1:20
    temp2=0;
    for k=1:1:length(p)
    temp2=temp2+((q(k)^n)/n)
    end
    ccoef2{i}(n)=temp2;
end
figure(2)
subplot(4,5,i)
plot(abs(ccoef1{i}),'b')
hold on
plot(abs(ccoef2{i}),'r')
end
```

3.8 Line Spectral Frequencies

Let the denominator term of the vocal tract filter be represented as $D(Z) = 1 - \sum_{k=1}^{k=r} a_k Z^{-k}$. The roots of the polynomial $D(Z)$ gives the poles of the vocal tract (x_k) and the transfer function can be represented as

$$D(Z) = \prod_{k=1}^{k=r}(1 - x_k Z^{-1}) = \prod_{k=1}^{k=r} \frac{(Z - x_k)}{Z} \tag{3.11}$$

Using $D(Z)$, the line spectral frequencies are obtained by computing the roots of the following polynomials.

$$L_1(Z) = D(Z) + Z^{-(r+1)} D^*(Z^{*-1})$$
$$L_2(Z) = D(Z) - Z^{-(r+1)} D^*(Z^{*-1})$$

The roots of the polynomial $L_1(Z)$ and $L_2(Z)$ occur when $L_1(Z) = 0$ and $L_2(Z) = 0$, respectively. Equating $L_1(Z) = 0$ and solving for Z gives the following.

$$D(Z) + Z^{-(r+1)} D^*(Z^{*-1}) = 0 \tag{3.12}$$

$$D(Z) = -Z^{-(r+1)} D^*(Z^{*-1}) \tag{3.13}$$

$$\frac{|D(Z)|}{|-Z^{-(r+1)} D^*(Z^{*-1})|} = 1 \tag{3.14}$$

$$\Rightarrow \frac{(D(Z))(D(Z))^*}{(-Z^{-(r+1)} D^*(Z^{*-1}))(-Z^{-(r+1)} D^*(Z^{*-1}))^*} = 1 \tag{3.15}$$

From (3.11), we get $D^*(Z^{-1*}) = \prod_{k=1}^{k=r} \frac{Z^* - x_k^*}{Z^*}$ at $Z = Z^{*-1}$. This implies,

$$D^*(Z^{-1*}) = \prod_{k=1}^{k=r} \frac{Z^{-1} - x_k^*}{Z^{-1}} \tag{3.16}$$

$$\Rightarrow D^*(Z^{-1*}) = \prod_{k=1}^{k=r} \frac{1 - x_k^* Z}{Z Z^{-1}} = \prod_{k=1}^{k=r} (1 - x_k^* Z) \tag{3.17}$$

Substituting (3.11) and (3.17) in (3.15), we get the following.

$$\frac{\prod_{k=1}^{k=r} \frac{(Z-x_k)(Z-x_k)^*}{|Z|^2}}{|Z|^{-2(r+1)} \prod_{k=1}^{k=r} (1 - x_k^* Z)(1 - x_k^* Z)^*} \tag{3.18}$$

$$\Rightarrow \frac{\prod_{k=1}^{k=r} (Z - x_k)(Z - x_k)^* |Z|^{2r}}{\prod_{k=1}^{k=r} (1 - x_k^* Z)(1 - x_k^* Z)^*} \tag{3.19}$$

The magnitude of the individual term of $\frac{\prod_{k=1}^{k=r}(Z-x_k)(Z-x_k)^*}{\prod_{k=1}^{k=r}(1-x_k^* Z)(1-x_k^* Z)^*}$ is equated to 1 to obtain the overall magnitude equals to 1. Let the kth term is equated to 1 and the following condition is achieved.

$$\frac{(Z - x_k)(Z - x_k)^*}{(1 - x_k^* Z)(1 - x_k^* Z)^*} = 1$$

$$\Rightarrow |Z|^2 + |x_k|^2 - Z x_k^* - Z^* x_k = 1 + |x_k|^2 |Z|^2 - x_k Z^* - x_k^* Z$$

$$\Rightarrow |Z|^2 + |x_k|^2 = 1 + |x_k|^2 |Z|^2$$

$$\Rightarrow |Z|^2 + |x_k|^2 - |x_k|^2 |Z|^2 = 1$$

$$\Rightarrow (|Z|^2 - 1)(|x_k|^2 - 1) = 0$$

If the magnitude of the x_k is less than 1, $(|x_k|^2 - 1) \neq 0$, and hence, $|Z|^2 - 1 = 0$. This in further implies $|Z| = 1$. Thus if all the poles lie within the unit circle, the magnitude of the individual term of $\frac{\prod_{k=1}^{k=r}(Z-x_k)(Z-x_k)^*}{\prod_{k=1}^{k=r}(1-x_k^* Z)(1-x_k^* Z)^*}$ is equal to 1. This in further implies that the magnitude of (3.19) is equal to 1. Thus, the roots of the $L_1(Z)$ lie on the unit circle. Similarly, it is found that the roots of the $L_2(Z)$ lie on the unit circle.

Let the roots of the $L_1(Z)$ be represented as $P_1, P_3, \ldots P_{r-1}$ and the roots of the polynomial $L_2(Z)$ be represented as $P_2, P_4, \ldots P_r$. It can be easily verified that $\angle(P_1) < \angle(P_2) < \angle(P_3) \cdots \angle(P_{r-1})\angle(P_r)$. This is known as interleaving. As the roots of the $L_1(Z)$ and $L_1(Z)$ lie on the unit circle, the solution is completely described by the phase at the unit circle. Also we understand that the roots occur when the polynomial $\frac{D(Z)}{Z^{-(r+1)}D^*(Z^{*-1})} = 1$ or $\frac{D(Z)}{Z^{-(r+1)}D^*(Z^{*-1})} = -1$. Let the polynomial $\frac{D(Z)}{Z^{-(r+1)}D^*(Z^{*-1})}$ is represented with $Z = e^{jw}$ to obtain $\frac{e^{-jwr}\prod_{k=1}^{k=r}(e^{jw}-x_k)}{e^{-iw(r+1)}\prod_{k=1}^{k=r}(1-x_k^* e^{jw})} = \frac{e^{-jwr}\prod_{k=1}^{k=r}(e^{jw}-x_k)}{e^{jwr}e^{-iw(r+1)}\prod_{k=1}^{k=r}(e^{-jw}-x_k^*)} = \frac{e^{jw(1-r)}\prod_{k=1}^{k=r}(e^{jw}-x_k)}{\prod_{k=1}^{k=r}(e^{-jw}-x_k^*)}$. Consider the argument of term $(e^{jw} - x_r)$. As w ranges from 0 to 2π, the change in the argument is obtained as 2π if x_r lies within the unit circle. Thus, the change in the argument of the term $\prod_{k=1}^{k=r}(e^{jw} - x_k)$ is obtained as $2 * \pi * r$ as w ranges from 0 to 2π. Also the change in the phase obtained by the term $\prod_{k=1}^{k=r}(e^{-jw} - x_k^*)$ is obtained as $-2 * \pi * r$ as w ranges from 0 to 2π. Similarly, the change in the phase obtained by the term $e^{jw(1-r)}$ is given as $-\pi(1-r)$ as w ranges from 0 to 2π. Thus, the overall phase difference achieved by the term $\frac{\prod_{k=1}^{k=r}(e^{jw}-x_k)}{e^{-iw(r+1)}\prod_{k=1}^{k=r}(1-x_k^* e^{jw})}$ is given as $2 * \pi * r + 2\pi(1-r) + 2 * \pi * r = 2\pi(r+1)$.When the phase response reaches π and 2π, the conditions $\frac{D(Z)}{-Z^{-(r+1)}D^*(Z^{*-1})} = -1$ and $\frac{D(Z)}{-Z^{-(r+1)}D^*(Z^{*-1})} = 1$ are satisfied, respectively. As w changes from 0 to 2π, the phase reaches π and 2π alternatively $(r+1)$ times. The ws corresponding the phase change of π are the roots of the $L_1(Z)$. Similarly, the ws corresponding the phase change of π are the roots of the $L_2(Z)$. From the discussion, we understand the roots of $L_1(Z)$ and $L_2(Z)$ are interleaved. The LSF of the two speech segments corresponding to the identical words are mentioned in Figs. 3.15 and 3.16, respectively. As the solution lies on the unit circle, the phase of the roots depict the roots and are plotted for the two identical words in Fig. 3.17 for illustration.

```
%linespectralfrequency.m
function [LSP1,LSP2]=linespectralfrequency(S1,S2,FS)
[res1,res2,S11,utforste,ltforste,ltforzcr]=endpointdetection(S1,FS);
[res1,res2,S12,utforste,ltforste,ltforzcr]=endpointdetection(S2,FS);
S11=dynamictimewarping1(S11,S12,FS);
lpccoef11=mainprogramforlpc(S11,FS);
lpccoef12=mainprogramforlpc(S12,FS);
j=1;
for i=1:1:size(lpccoef11,2)
R=roots(lpccoef11(:,i));
R1=1./conj(R);
R2=poly(R1);
R3=[zeros(1,13) 1];
R4=conv(R2,R3);
R5=[zeros(1,13) -1];
R6=conv(R2,R5);
lpccoef11new(:,i)=[lpccoef11(:,i)' zeros(1,13) ];
poly1(:,i)=lpccoef11new(:,i)+R4';
linespectrum1(:,i)=roots(poly1(:,i));
poly2(:,i)=lpccoef11new(:,i)+R6';
linespectrum2(:,i)=roots(poly2(:,i));
```

```
figure(1)
subplot(3,6,j)
hold on
zplane([],[linespectrum1(:,i) linespectrum2(:,i)])
figure(2)
subplot(4,5,j)
plot(angle(linespectrum1(:,i)),'bo')
hold on
plot(angle(linespectrum2(:,i)),'bo')
LSP1{i}=[linespectrum1(:,i) linespectrum2(:,i)];

R=roots(lpccoef12(:,i));
R1=1./conj(R);
R2=poly(R1);
R3=[zeros(1,13) 1];
R4=conv(R2,R3);
R5=[zeros(1,13) -1];
R6=conv(R2,R5);
lpccoef12new(:,i)=[lpccoef12(:,i)' zeros(1,13) ];
poly1(:,i)=lpccoef12new(:,i)+R4';
linespectrum1(:,i)=roots(poly1(:,i));
poly2(:,i)=lpccoef12new(:,i)+R6';
linespectrum2(:,i)=roots(poly2(:,i));
figure(3)
subplot(3,6,j)
hold on
zplane([],[linespectrum1(:,i) linespectrum2(:,i)])
figure(2)
subplot(4,5,j)
plot(angle(linespectrum1(:,i)),'r*')
```

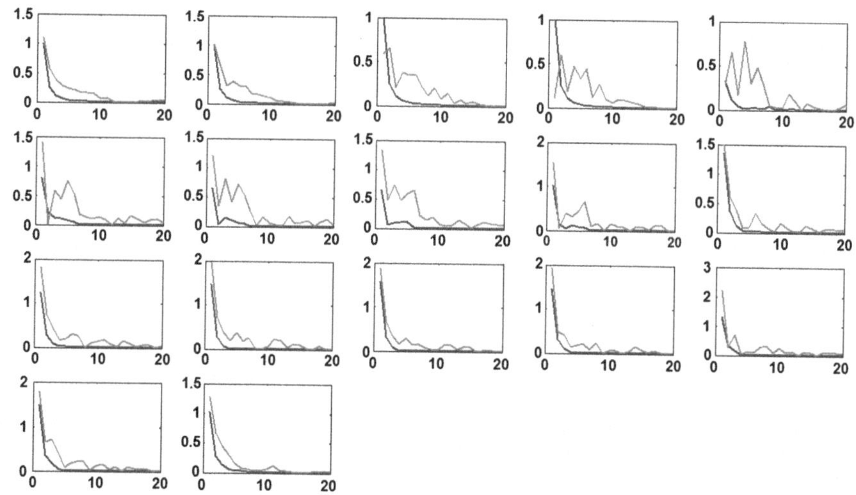

Fig. 3.14 Absolute values of the cepstral co-efficients for $n = 1$ to $n = 20$ of the individual frames of the two speech segments (corresponding to the identical words mentioned in Fig. 3.9)

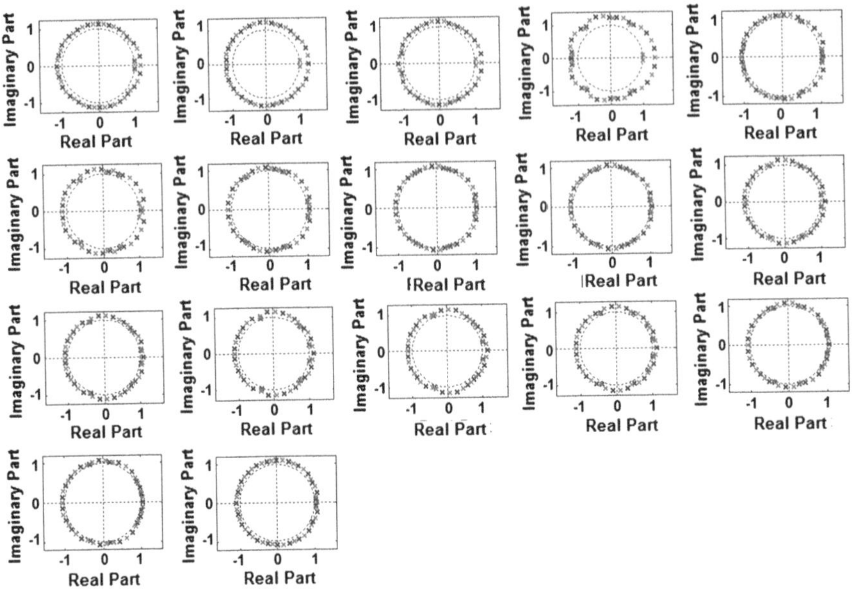

Fig. 3.15 Line spectral frequencies of the individual frames of the speech segment mentioned in the subplot 1 of Fig 3.9. The interleaving is seen in the plot. It is also seen that the magnitude of the line spectrum is almost near to 1

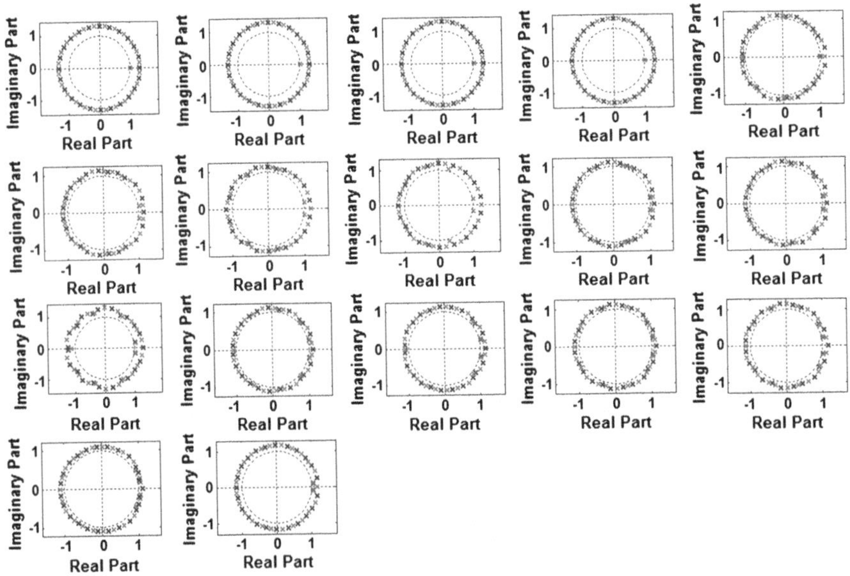

Fig. 3.16 Line spectral frequencies of the individual frames of the speech segment mentioned in the subplot 2 (corresponding to the identical word mentioned in Fig. 3.15) of Fig 3.9. It is seen that the plot is identical with Fig. 3.15

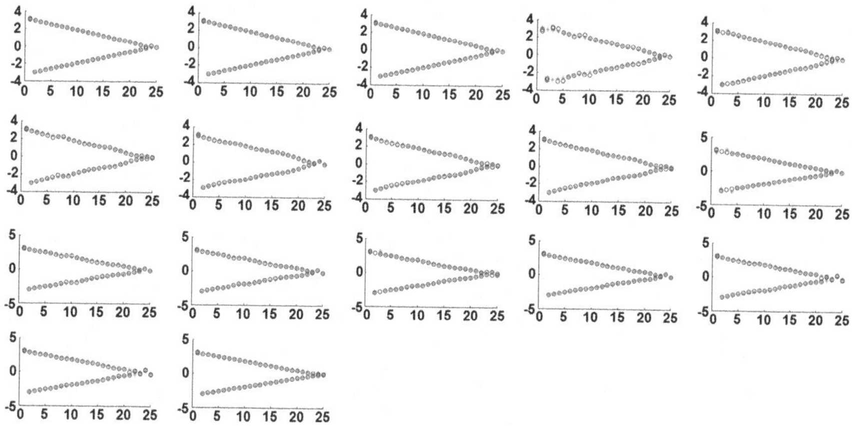

Fig. 3.17 Angle (*in radians*) of the LSF of the individual frames of the speech segment of the individual frames of the two speech segments (corresponding to the identical words mentioned in Fig. 3.9)

```
hold on
plot(angle(linespectrum2(:,i)),'r*')
LSP2{i}=[linespectrum1(:,i) linespectrum2(:,i)];
j=j+1;
end
```

Consider all-pole stable filter (poles lie within the unit circle) with transfer function given as

$$H(Z) = \frac{\sum_{i=1}^{i=4}(Z - r_i)}{Z^{-1} - r_5^*} \tag{3.20}$$

with $r_1 = 0.5 + 0.5j, r2 = 0.3 + 0.2j, r3 = 0.4 - 0.3j, r4 = 0.1 + 0.3j, r5 = 0.3 + 0.5j$. The phase response of the filter crosses π and $-\pi$, 5 times as expected as shown in Fig. 3.18.

3.9 Mel-Frequency Cepstral Co-efficients

The human perceptual frequency is represented in mel scale and it is related to the actual frequency as $2595 \log(1 + \frac{f}{700})$ (refer Fig. 3.19). The human perception of sound is assumed to be consist of bank of filters (cochlea model (refer Appendix F)). Each filter is of triangular in shape. The triangular filter banks in mel scale are uniformly spaced. Accordingly, the bandwidth of the individual filter increases logarithmically in the normal scale (refer Fig. 3.20). Each triangular filter is of length 1,000 (arbitrarily chosen) in frequency domain. Also note that the 1,000th sample corresponds to $\frac{F_s}{2}$. Let the magnitude response of the ith triangular filter is

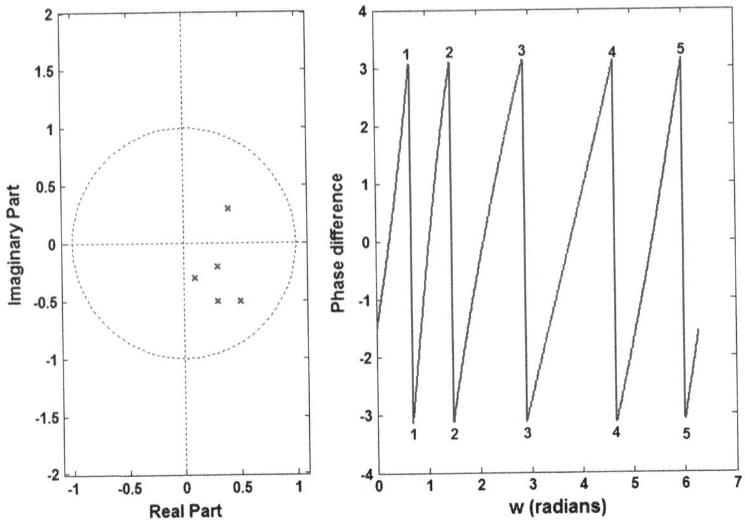

Fig. 3.18 Illustration of number of times (5 times) the phase reaches π or $-\pi$ of the all-pole stable filter with transfer function given in (3.19)

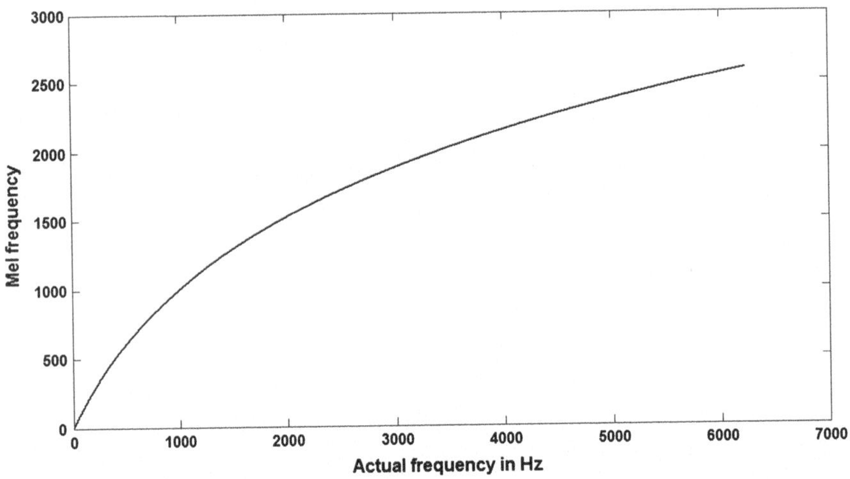

Fig. 3.19 Illustration of the graph relating actual frequencies with the mel scale

represented as the column vector H_i. Also let the number of filters used is 24. The MFCC is obtained for every 30 ms speech segment (with overlapping of 10 ms) of the particular word for isolated speech/speaker recognition. The steps involved in extracting MFCC for every segment is described below.

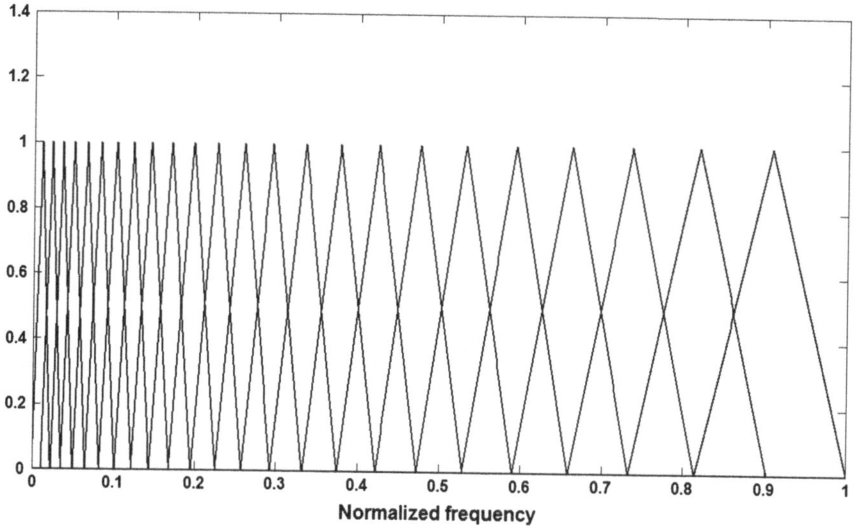

Fig. 3.20 Bank of filters in mel-frequency scale. 1 in x-axis corresponds to $\frac{F_s}{2}$

1. Let the speech segment is represented as the column vector S. The speech segment S is multiplied with the Hamming window to avoid Gibbs phenomenon (refer Sect. 3.9.1) to obtain the preprocessed speech segment S_1. Compute the 1998—*point* Fast Fourier Transformation (FFT) of the preprocessed speech segment. Collect the absolute values of the first 1,000 samples. Let it be represented as the column vector S_2. Note that 1, 000th sample corresponds to $\frac{F_s}{2}$.
2. Compute the energy of the speech segment as $\log_{10}(S_2^T S_2)$. This is treated as the first element of the MFCC. Let it be m_1.
3. Compute $\log_{10}((S_2^T H_i)^2)\ \forall i = 1 \ldots 24$ to obtain 24-dimensional column vector S_3.
4. Represent the vector S_3 as the linear combinations of discrete cosine basis (refer Sect. 3.9.2 to obtain 24 co-efficients.
5. First coefficient corresponds to DC component, and hence, it is not considered. The next 12 co-efficients are collected to obtain m_2 to m_{13}. Thus, 13-dimensional MFCC vector $M = [m_1\ m_2\ \ldots m_{13}]$ is obtained.
6. The first derivative of the MFCC (excluding the first coefficient) is computed as the difference between the adjacent samples of the vector M to obtain 11 co-efficients vector M'. The difference between the adjacent samples of the vector M' is computed as the second-derivative MFCC to obtain 10 co-efficients vector M''.
7. Thus, $1 + 12 + 11 + 10 = 34$ dimensional MFCC is obtained for the particular speech segment. Two speech segments corresponding to the identical isolated words and the corresponding MFCC are illustrated in Fig. 3.21.

```
%collectmfcc.m
function [mfcoef]=collectmfcc(x,FS)
f=1:1:FS/2;
mel=2595*log10(1+f/700);
figure(1)
plot(f,mel)
%Uniform in mel frequency
s=mel(FS/2)/25;
%Identify the center frequency in normal scale.
COL=[];
for i=1:1:25
[p,r]=min(abs(mel-i*s));
COL=[COL f(r)];
end
COL=[0 COL]/max(COL);
figure(2)
%Create bank of filters
xrange=([0:1:999]/999);
for i=1:1:24
    x1=COL(i):0.001:(COL(i+2)+COL(i))/2;
    y1=(2/(COL(i+2)-COL(i)))*x1-(2/(COL(i+2)-COL(i)))*COL(i);
    x2=(COL(i+2)+COL(i))/2:0.001:COL(i+2);
    y2=(-2/(COL(i+2)-COL(i)))*x2+(2/(COL(i+2)-COL(i))*COL(i+2));
    a=zeros(1,1000);
    a(round(x1*999+1))=y1;
    a(round(x2*999+1))=y2;
    filter_col{i}=a;
    plot(xrange,a)
    hold on
end
sbs=fix(FS*25*10^(-3));
ovs=fix(FS*10*10^(-3));
mfcoef=blkproc(x,[1 sbs],[0 ovs],'obtainmfcc(x,P1)',filter_col);

%obtainmfcc.m
function [res]=obtainmfcc(x,filter_col)
%Apply hammingwindow and followed by fft
y=hamming(length(x))'.*x;
x1=abs(fft(y,1998));
x2=x1(1:1:1000);
res1=log10(sum(abs(fft(y)).^2));
res2=[];
for i=1:1:24
res2=[res2 sum((x2.*filter_col{i}).^2)];
end
res3=log10(res2);
%Suppose the energy values are scaled by some constant number,
%taking log
%will have the corresponding DC shift. This is further removed
%using dct as follows.
res4=dct(res3);
```

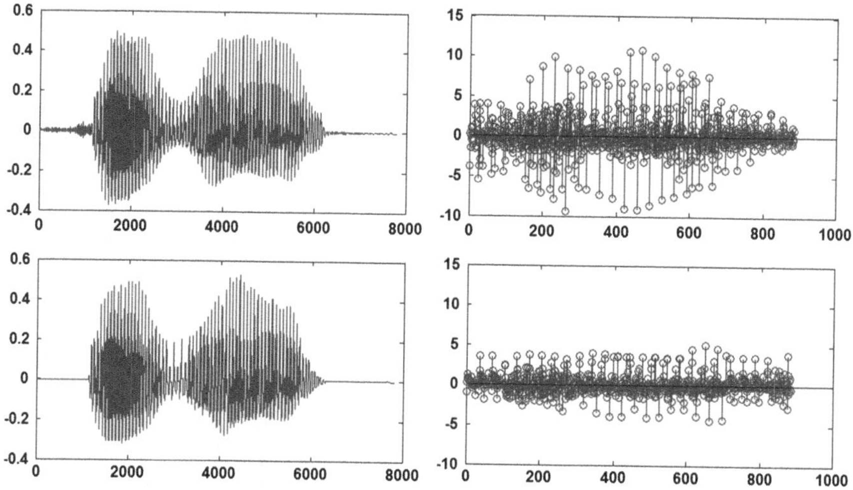

Fig. 3.21 Two MFCC corresponding to two isolated speech segments corresponding to the identical isolated words

```
res5=res4(2:1:13);
res6=diff(res5);
res7=diff(res6)
res=[res1 res5 res6 res7];
```

3.9.1 Gibbs Phenomenon

Let the FT of the speech signal $s(t)$ is represented as $S(f)$. Let the speech signal $s(t)$ is truncated for the finite duration (0 to T) and is represented as $s(t)w(t)$, where $w(t) = 1 \; \forall \, 0 \le t \le T$ and 0 elsewhere. The FT of the truncated signal $s(t)w(t)$ is computed as the convolution of $S(f)$ with $W(f)$ where $W(f)$ is the FT of the rectangular window. The magnitude response of the $W(f)$ is represented in Fig. 3.22. This lead to the introduction of more ripples in the magnitude response of the FT of the truncated signal $s(t)w(t)$ as shown in Fig. 3.23 . This is known as Gibbs phenomenon. This is circumvented by multiplying the speech signal $s(t)$ with the Hamming or other smoothing window $h(t)$ (refer (3.22)) as $s(t)h(t)$. The Hamming window in time domain is given in Fig. 3.24. Thus, the magnitude response of the FT of the $s(t)h(t)$ will not have ripples as compared to the ripples present in the FT of $s(t)w(t)$ (refer Fig. 3.23).

$$W(f) = Te^{-j2\pi fT} \frac{sin\pi fT}{\pi fT} \tag{3.21}$$

$$h(t) = 0.54 - 0.46\,cos\left(\frac{2\pi t}{T}\right) \tag{3.22}$$

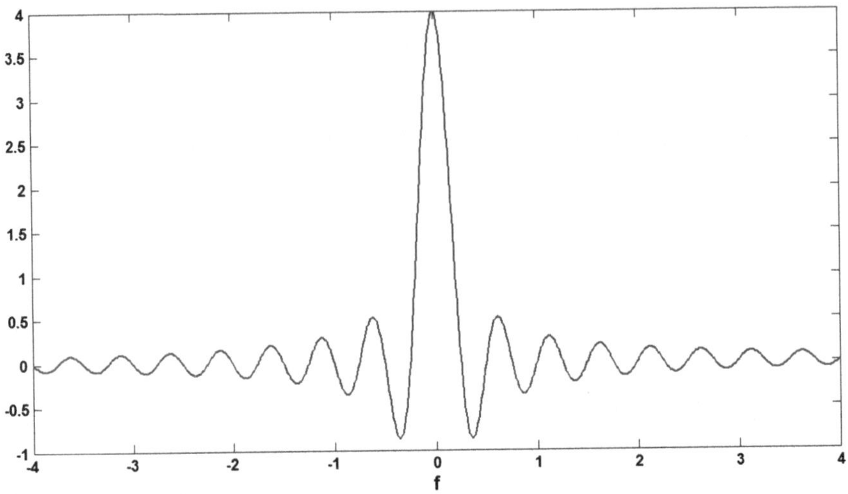

Fig. 3.22 Magnitude response of the FT of the rectangular window that leads to gibbs phenomenon

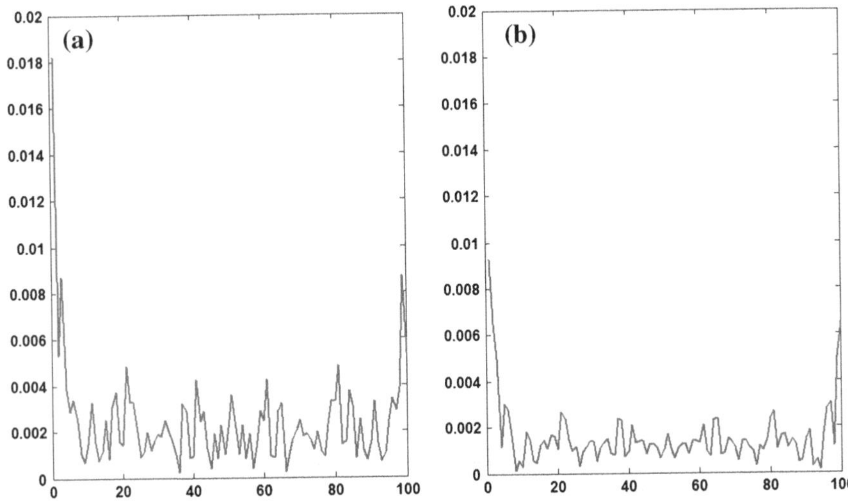

Fig. 3.23 Magnitude response of the FFT of the typical speech signal with (**a**) rectangular window (**b**) Hamming window

3.9.2 Discrete Cosine Transformation

The discrete cosine transformation (DCT) equation is as given below.

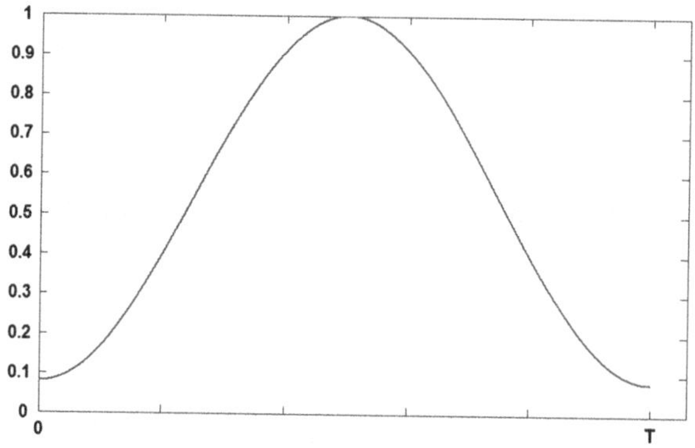

Fig. 3.24 Hamming window in time domain

$$X(k) = \frac{2}{\sqrt(N)} \sum_{n=1}^{n=N} x(n)\, cos\left(\frac{\pi(2n-1)(k-1)}{2N}\right) \tag{3.23}$$

$\forall k = 2, \ldots N$

$$X(1) = \frac{1}{\sqrt(N)} \sum_{n=1}^{n=N} x(n) \tag{3.24}$$

Consider the vector x with size $N \times 1$ and the corresponding DCT vector computed using (3.23) and (3.24) be represented as X. The elements of the vector X can be viewed as the co-efficients to represent the vector x as the linear combinations of the DCT basis vectors. The DCT bases vectors are obtained as the DCT of the standard bases vectors and are represented in Fig. 3.25 for $N = 24$.

```
%dctbasisdemo.m
%DCT bases plot for N=24
a=[1 zeros(1,23)]';
col=dct(a)';
b=dct(a');
figure
subplot(6,4,1)
plot(col(:,1))
i=1;
for j=2:1:24
subplot(6,4,i)
plot(b)
a=circshift(a,1);
b=dct(a);
i=i+1;
col=[col;b'];
```

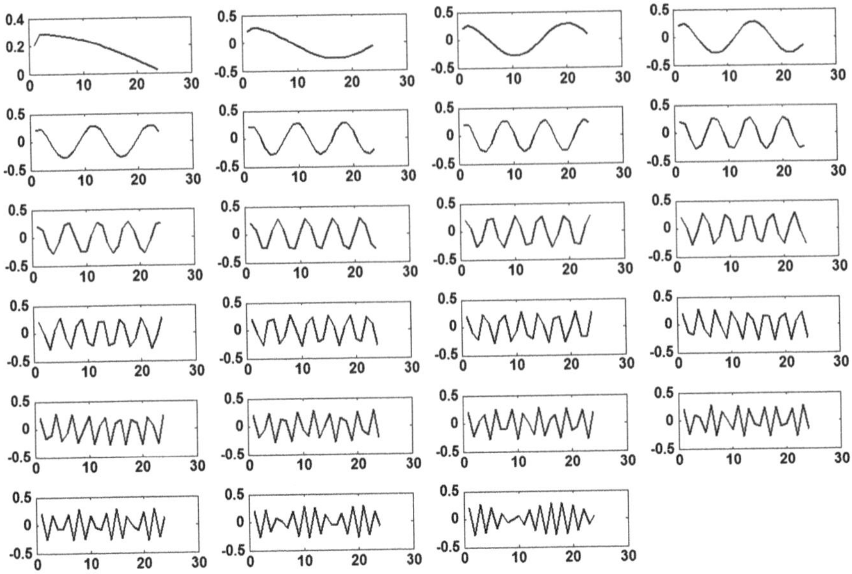

Fig. 3.25 Basis of the DCT for $N = 24$

```
subplot(6,4,i)
plot(col(:,i))
end
```

3.10 Spectrogram

Consider the two signals $s_1(t) = A_1\,sin(2\pi f_1 t) + A_2\,sin(2\pi f_2 t) + A_3\,sin(2\pi f_{32} t)$ $\forall\ 0 < t < T_3$ and $s_2(t) =$

$$A_1 sin(2\pi f_1 t) \forall\ 0 < t < T_1$$
$$A_2 sin(2\pi f_2 t) \forall\ T_1 < t < T_2$$
$$A_3 sin(2\pi f_3 t) \forall\ T_2 < t < T_3$$

The signal $s_1(t)$ is having frequency f_1, f_2 and f_3 for the complete time duration. But signal $s_2(t)$ is having f_1 for $0 < t < T_1$, f_2 for $T_1 < t < T_2$ and f_3 for $T_2 < t < T_3$. The spectrum (magnitude response of the FT of the signal) of the two signals $s_1(t)$ and $s_2(t)$ will have peaks at f_1, f_2 and f_3. So the spectrum is not able to distinguish the two signals (refer Fig. 3.26). This is circumvented using short-time Fourier transformation (STFT). This is obtained by computing the FT of the signal multiplied with the sliding window. If the window is small, time resolution is increased. But it may not accommodate certain lower frequencies, and hence, frequency resolution is decreased. If the window is large, time resolution is decreased,

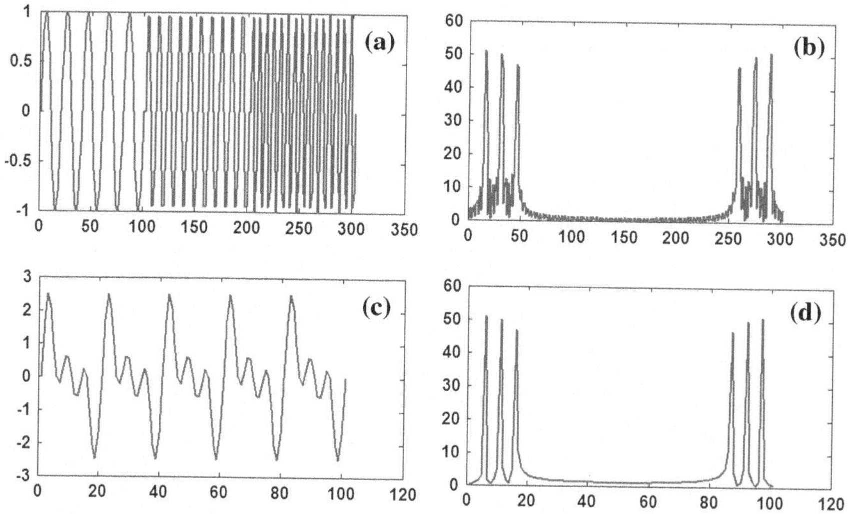

Fig. 3.26 **a** Discrete samples of $s_1(t)$ (refer Sect. 3.8) (i.e., $s_1(n)$). **b** Magnitude response of DFT of $s_1(n)$. **c** $s_2(n)$. **d** Magnitude response of DFT of $s_2(n)$

but frequency resolution is increased. $X(\tau, f) = \int_{-\infty}^{\infty} x(t)w(t-\tau)e^{-j2\pi ft}$. The discrete version of the STFT is the spectrogram as described below. $X(k,m) = \sum_0^{N-1} x(n-m)w(n)e^{\frac{-j2\pi(n-m)k}{N}}$ where $k = 0 \ldots N-1$. The magnitude-squared value $X(k,m)$ (i.e., $|X(k,m)|^2$) is computed to form the spectrogram as described below.

1. Collect the speech segment x for $n = 0$ to $n = N-1$. Multiply with the window. Compute N-point FFT and collect squared values for $k = 0 \cdots N/2$ (assuming N is even). This corresponds to $m = 0$.
2. Collect the speech segment x for $n = 1 \ldots N$. Multiply with the window. Compute N-point FFT and collect squared values for $k = 0 \ldots N/2$ (assuming N is even). This corresponds to $m = 1$.
3. In general, collect the speech segment x for $n = r \ldots = N-1+r$. Multiply with the window. Compute N-point FFT and collect squared values for $k = 0\, to\, N/2$ (assuming N is even). This corresponds to $m = r$.
4. Plot the graph with m in the x-axis and k in the y-axis and $X(k,m)$ as the color shades proportional to the $\log_{10}$ of the squared magnitude.
5. Most commonly used windows are the following. Hanning window (w_{han}), Hamming window (w_{ham}), and Blackmann windows (w_{black1} *and* w_{black2}) (refer Fig. 3.27). They are used to circumvent Gibbs phenomenon.

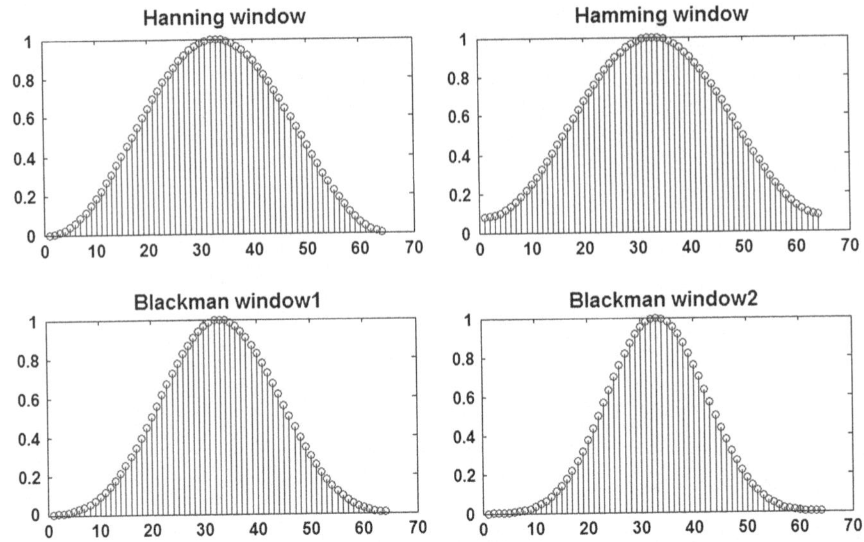

Fig. 3.27 Commonly used windows for $N = 64$

$$w_{han} = 0.5 - 0.5\,cos\left(\frac{2\pi n}{N}\right)$$

$$w_{ham} = 0.54 - 0.46\,cos\left(\frac{2\pi n}{N}\right)$$

$$w_{black1} = 0.423 - 0.498\,cos\left(\frac{2\pi n}{N}\right) + 0.079\,cos\left(\frac{4\pi n}{N}\right)$$

$$w_{black2} = 0.359 - 0.488\,cos\left(\frac{2\pi n}{N}\right) + 0.141\,cos\left(\frac{4\pi n}{N}\right) - 0.012\,cos\left(\frac{6\pi n}{N}\right)$$

The typical speech segment and the corresponding spectrogram for the window size of 15, 20, and 25 ms are plotted in Fig. 3.28. 512-point FFT is computed in all the cases by padding the required number of zeros for comparison. The spectrogram for constant $m = 3,000$ is plotted as the function of k, for different window sizes, namely 15, 20, and 25 ms (refer Fig. 3.29). The graph depicts that the higher frequency resolution is achieved for the window size $=$ 25 ms. Similarly, spectrogram for $k = 125$ is plotted as the function of m, for different window sizes, namely 15, 20, and 25 ms (refer Fig. 3.29). The graph depicts that better time resolution is obtained for the window size $=$ 15 ms.

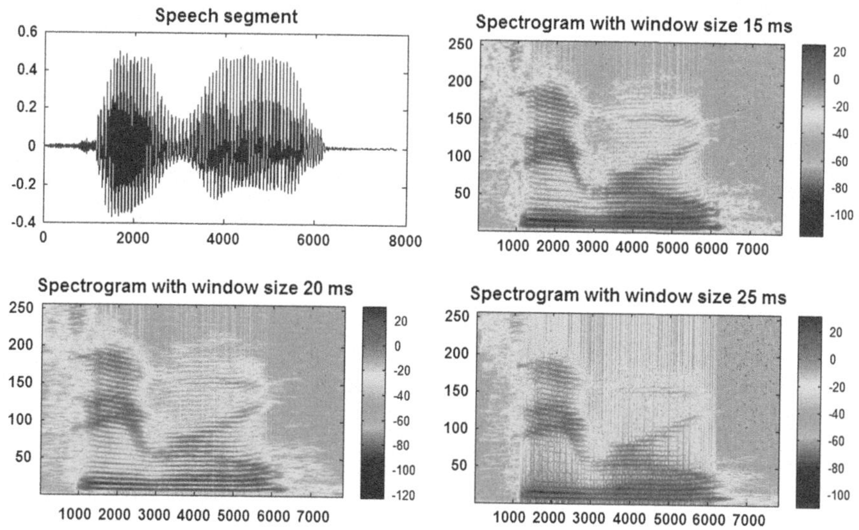

Fig. 3.28 Spectrograms computed for the three different window sizes

```
%spectrogram.m
function [SG1,SG2,SG3]=spectrogram(speechsegment,FS)
%This program uses the user-defined function sg.m
n1=nextpow2(FS*10*10^(-3));
N1=2^n1;
n2=nextpow2(FS*20*10^(-3));
N2=2^n2;
n3=nextpow2(FS*40*10^(-3));
N3=2^n3;
N=max([N1 N2 N3]);
SG1=blkproc(speechsegment,[1 1],[0 N1-1], 'sg(x,P1)',N3);
SG2=blkproc(speechsegment,[1 1],[0 N2-1], 'sg(x,P1)',N3);
SG3=blkproc(speechsegment,[1 1],[0 N3-1], 'sg(x,P1)',N3);
figure
subplot(2,2,1)
plot(speechsegment)
subplot(2,2,2)
imagesc(10*log10(SG1))
title('Spectrogram with window size 15 ms')
subplot(2,2,3)
imagesc(10*log10(SG2))
title('Spectrogram with windowsize 20 ms')
subplot(2,2,4)
imagesc(10*log10(SG3))
title('Spectrogram with window size 25 ms')

%sg.m
function [res]=sg(x,N)
x=x.*hamming(length(x))';
y=abs(fft(x,N)).^2;
res=y(1:1:N/2)';
```

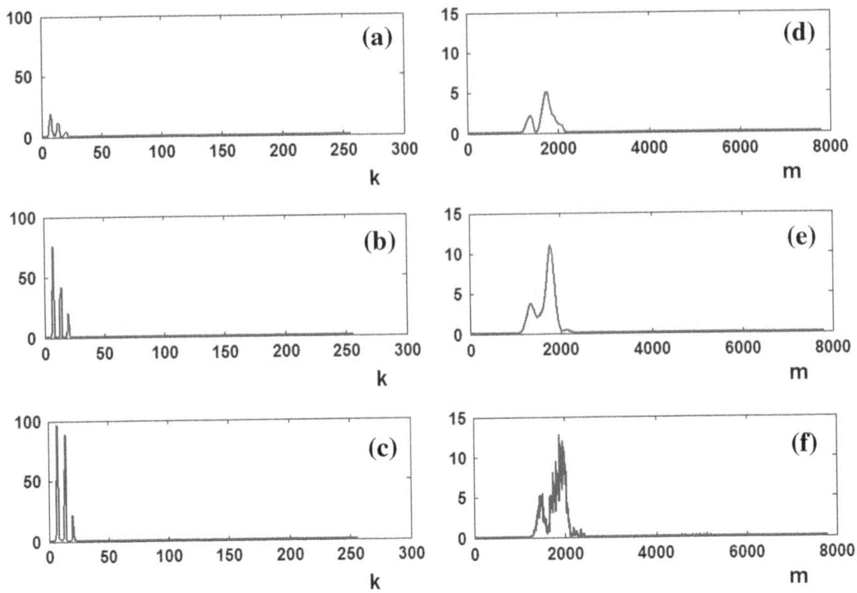

Fig. 3.29 Spectrogram for $m = 3000$ as the function of k with window size. (**a**) 15 ms (**b**) 20 ms (**c**) 25 ms (**d**) Spectrogram for $k = 125$ as the function of m with window size. (**d**) 15 ms (**e**) 20 ms (**f**) 25 ms

3.10.1 *Time Resolution Versus Frequency Resolution in Spectrogram*

The resolution of the time domain is determined by the sampling time. If it is small, the signal is having high time resolution. Let the window size be fixed as N samples. The N-point DFT is computed for the particular window. The resolution in frequency domain is determined as $\frac{FS}{N}$. Thus, if TS is small, FS becomes larger, and hence, $\frac{FS}{N}$ becomes larger. This indicate if the time resolution is increased (decreasing TS), the frequency resolution decreases. Not that using N-points in time domain, it is possible to obtain $M > N$ point DFT by padding zeros with the input signals. We get the DFT with M-point with $\frac{FS}{M}$ interval. This does not mean that the resolution has been increased. This helps to view the DFT data better. But it does not increase the resolution.

3.11 Discrete Wavelet Transformation

Discrete wavelet transformation (DWT) consists of four filters, namely low-pass decomposition filter (LDF), high-pass decomposition filter (HDF), low-pass

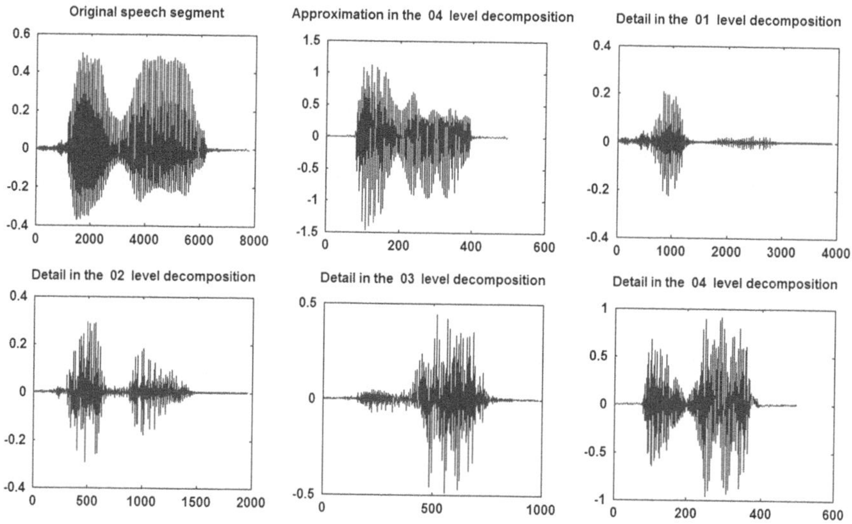

Fig. 3.30 Four-level decomposition of the speech signal using *db*6 wavelet

reconstruction filter (LRF), and high-pass reconstruction filter (HRF). The cutoff frequencies of all the filters are $\frac{\pi}{2}$. The speech signal S is passed through LDF and HDF and down sampled by 2 to obtain the $approx_1$ (approximation) and the $detail_1$ signals, respectively. This is the first-level wavelet decomposition. The $approx_1$ is treated as the original speech signal and is decomposed to obtain $approx_2$ and $detail_2$ signals using the corresponding decomposition filters. These are known as second-level wavelet decomposition. The $approx_n$ and all the detail co-efficients form the wavelet transformation of the nth-level wavelet decomposition. The $approx_n$ is upsampled to form the length equal to the number of the original speech signal and is passed through the LRF to form $approx_n R$. Similarly, the detail signals $detail_1, detail_2, detail_3 \ldots detail_n$ are upsampled individually to form the length equal to the number of the original speech signal and are passed through HRF to obtain $detail_1 R, detail_2 R, detail_3 R \ldots detail_n R$, respectively. Adding the signals $approx_n R + detail_1 R + detail_2 R + \cdots + detail_n R$ gives the original speech signal S. Based on the combinations of the filters used, the wavelet transformation is described as *Daubechies*, *Coiflets*, *Symlets*, *biorthogonal*, and *reverse biorthogonal*. The wavelet decomposition of the typical speech signal using *Daubechies* wavelet (db6) is shown in Fig. 3.30. Approximation in the fourth-level decomposition ($approx_4$) will have frequency ranging from 0 to $\frac{F_s}{32}$. Similarly, the $detail_1$ will have frequency ranging from $\frac{F_s}{4}$ to $\frac{F_s}{2}$, $detail_2$ will have frequency ranging from $\frac{F_s}{8}$ to $\frac{F_s}{4}$, $detail_3$ will have frequency ranging from $\frac{F_s}{16}$ to $\frac{F_s}{8}$ and $detail_4$ will have frequency ranging from $\frac{F_s}{32}$ to $\frac{F_s}{16}$

```
%wavtrans.m
function [approx detail]=wavtrans(speechsegment,N)
%'N' level Discrete wavelet transformation using 'db6' wavelet.
[LD ,HD,LR,HR]=wfilters('db6')
s=speechsegment;
for i=1:1:N
approx=conv(s,LD);
approx=dyaddown(approx,2);
detail=conv(s,HD);
detail=dyaddown(detail,2);
detailcoef{i}=detail;
s=approx;
end
m=round(sqrt(N+2));
n=round((N+2)/m);
figure
subplot(m,n,1)
plot(speechsegment)
title('Original speech segment')
subplot(m,n,2)
plot(approx)
title(strcat('Approximation in the  0', num2str(N),
              'level decomposition'))
for i=1:1:N
subplot(m,n,i+2)
plot(detailcoef{i})
title(strcat('Detail in the  0',num2str(i),'level decomposition'))
end
```

3.12 Pitch Frequency Estimation

3.12.1 Autocorrelation Approach

The voiced speech segment is quasi-periodic in nature (refer Fig. 3.31). The autocorrelation of the speech signal helps in estimating the pitch frequency as described below. Let S be the isolated voiced speech segment. The autocorrelation of the deterministic speech segment is computed as follows.

$$A(k) = \frac{1}{N}\sum_{n=1}^{n=N} S(n)S(n-k) \tag{3.25}$$

If the pitch frequency is f_p for the discrete isolated voiced speech segment with sampling frequency f_s, pitch period in samples is calculated as $\frac{f_s}{f_p}$. Usually, the pitch frequency ranges from 85 to 180 Hz for male and 165 to 255 Hz for female. The pitch frequency ranges is from 49 to 147 Hz. The corresponding pitch period for the 12500 Hz sampling frequency speech segment is computed as 49 to 147 samples.

Thus, $A(k)$ is computed for $k = 44$ to 133. The k corresponding to the maximum value of $A(k)$ is declared as the pitch period in samples p(say), and the corresponding pitch frequency is calculated as $\frac{f_s}{f_p}$.

```
%pitchestautocorr.m
function [pitchfrequency]=pitchestautocorr(S,FS)
[res1,res2,speechsegment,utforste,ltforste,ltforzcr]=endpointdetection(S,FS)
SEGLEN=FS*30*10^(-3);
S=speechsegment;
S1=S(fix(length(S)/2):SEGLEN-1+fix(length(S)/2));
T=[];
HIGH=fix(FS/85);
LOW=fix(FS/255);
for k=LOW:1:HIGH
     T=[T  sum([zeros(1,k) S1 ].*[S1 zeros(1,k) ])];
end
[P,Q]=max(T);
pitchperiod=Q+LOW;
pitchfrequency=FS/pitchperiod;
```

3.12.2 Homomorphic Filtering Approach

In the source-filter model of the speech production, excitation source is either noise or impulse stream. The excitation source is passed through the vocal tract filter to obtain the speech signal. Let the excitation source, vocal tract filter impulse response, and the corresponding speech signal are represented as $E(Z)$, $V(Z)$, and $S(Z)$, respectively, in the Z-domain. They are related as $S(Z) = E(Z)V(Z)$. The pitch of the speech signal is the frequency of the impulse stream (excitation source for voiced speech segment). Hence, we need to separate $E(Z)$ from $S(Z)$ to obtain the pitch frequency. This is obtained as follows. Take log on $S(Z)$ to obtain $\log_{10} S(Z) = \log_{10} E(Z)V(Z) = \log_{10} E(Z) + \log_{10} V(Z)$ to convert the multiplication into addition. $E(Z)$ is boosted in this case. Compute the time-domain equivalent of $\log_{10} E(Z) + \log_{10} V(Z) =$ to estimate pitch frequency. This is achieved using the following steps.

1. Multiply the particular isolated speech segment with the Hamming window to reduce Gibbs phenomenon. Let it be S_1.
2. Compute the absolute of the FFT of the particular speech segment. Let it be S_2.
3. Compute the $20 * \log_e(S_2)$. Let it be S_3.
4. Compute the absolute of the IFFT of the S_3. Let it be S_4.
5. Smooth the signal S_4. Let it be S_5.
6. Identify the position of the maximum amplitude of S_5 among the first half of the samples, other than the region around the first sample. Let it be N.
7. The pitch frequency is calculated as $\frac{FS}{N}$.

The pitch frequency computed for two speech segments are illustrated in Fig. 3.32 and the corresponding pitch frequencies are computed as 235.85 and 162.34 Hz, respectively.

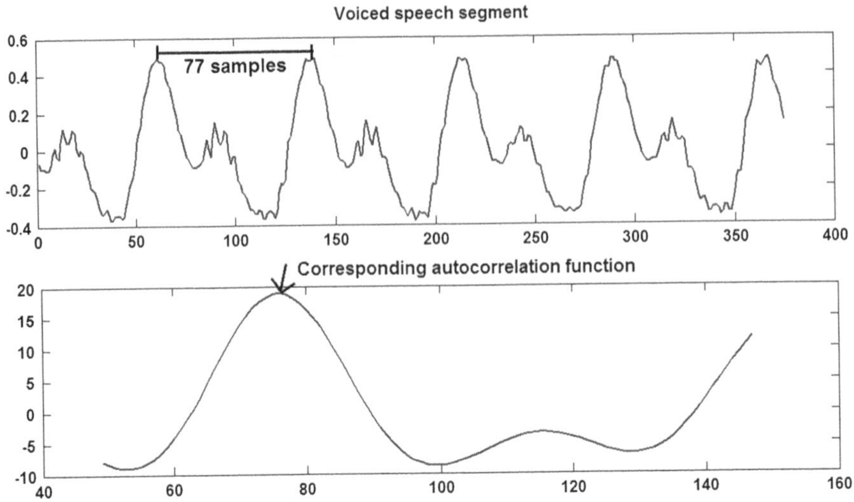

Fig. 3.31 Typical voiced speech segment and the corresponding autocorrelation function. The arrow mark indicates the pitch period

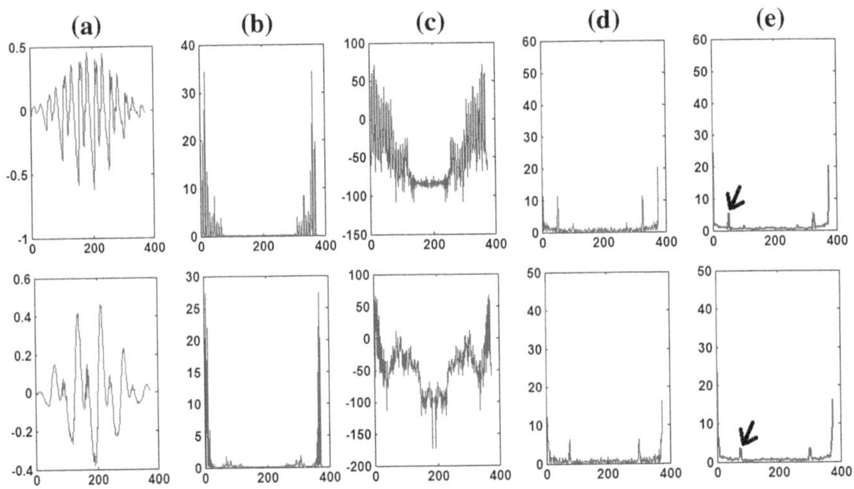

Fig. 3.32 Illustration of the pitch frequency estimation using homomorphic filtering. **a** Speech segment. **b** Corresponding magnitude response of the speech signal. **c** Log of the corresponding magnitude response of the speech signal. **d** absolute value of the IFFT of the signal mentioned in *c*. **e** Signal in *c* after smoothing. The arrow mark indicates the pitch frequencies

```
%pitchfreqest.m
function [PF]=pitchfreqest(S,FS);
[res1,res2,speechsegment,utforste,ltforste,ltforzcr]=endpointdetection(S,FS)
S=speechsegment;
i=1;
```

```
%Identify the frame for extracting pitch
S1=S(fix(length(S)/2):374+fix(length(S)/2));
S1=S1.*hamming(length(S1))';
subplot(2,5,i)
plot(S1)
subplot(2,5,i+1)
S2=abs(fft(S1));
plot(S2)
S3=20*log(S2);
subplot(2,5,i+2)
plot(S3)
S4=abs(ifft(S3));
subplot(2,5,i+3)
plot(S4)
subplot(2,5,i+4)
S5=smooth(S4,5);
plot(S5)
S5=S5(1:1:188);
[p,q]=max(S5(9:1:length(S5)));
PF=FS/(q+9);
```

3.13 Formant Frequency Estimation

3.13.1 Formant Extraction Using Vocal Tract Model

The resonant frequencies of the vocal tract transfer function $V(Z)$ (refer (3.1)) are known as formant frequencies. Usually, there are three dominant resonant frequencies. They are known as formant-1 (270–730 Hz), formant-2 (840–2290 Hz), and formant-3 (1690–3010 Hz) frequencies. The vocal tract is modeled using LPC as discussed in the Sect. 3.3 and the formant frequencies are estimated by identifying the peak around the formant-1, formant-2, and formant-3 frequencies as demonstrated in Fig. 3.33. The 3-dB bandwidth of the resonant frequency does not increase more than 160, 200, and 300 Hz for formant-1, formant-2, and formant-3, respectively. The peak corresponds to the poles of the vocal tract filter. Hence, roots of the polynomial $A(Z) = \frac{1}{V(Z)}$, that have the bandwidth less than 160, 200, 300 Hz are estimated as the formant-1, formant-2, and formant-3 frequencies, respectively. The formant frequencies are estimated as follows.

1. Multiply the speech segment with the Hamming window.
2. Estimate 11th-order LPC and $A(Z)$ using the technique described in the Sect. 3.3.
3. Compute the roots of the polynomial $A(Z)$.
4. Compute the phase of the roots ranging from 0 to π. Let it be $\omega_1 \ldots \omega_n$.
5. Compute the 512-point FFT of the $A(Z)$ and its normalized absolute values are represented as H (maximum amplitude of 1). Note the magnitude of H corresponding to $\omega_1 \ldots \omega_n$ is obtained as $H(m_1) \ldots H(m_n)$ where $m_i = round(\frac{\omega_i * 512}{\pi})$.

6. The maximum $\frac{bandwidth}{2}$ at the formant frequencies does not exceed 160, 200, and 300 Hz, respectively, for formant-1, formant-2 and formant-3. For formant-1, this is computed in terms of number of samples as $N = round(\frac{80*1024}{FS})$. For $FS = 12{,}500$, the value of N is computed as $N = 7$.
7. Collect the values of H for the range $m_1 - 2 * N$ to $m_1 + 2 * N$ as the vector H_1. If the maximum value of H_1 is greater than 0.01, compute the index of the maximum value. Let it be p_1. Compute the ratio $R_1 = \frac{H_1(p_1)}{H_1(p_1-N)}$. If the ratio R_1 is less than the threshold $th = 1.414$, declare $omega_1$ as the formant frequency.
8. Repeat the step 7 for m_2 to m_n to obtain the remaining formant frequencies (if available).
9. If $p_1 < N$, repeat the step 7 with $N = 1$ with the threshold $th = \frac{1.414}{N}$.

```
%endpointdetection.m
[res1,res2,speechsegment,utforste,ltforste,ltforzcr]
    =endpointdetection(S,FS)
%->F1 270 to 730  F2-840 to 2290 F3-1690 to 3010 Hz
%160 HZ 200 Hz and 300 Hz
SEGLEN=FS*30*10^(-3);
S=speechsegment;
S1=S(fix(length(S)/2):SEGLEN-1+fix(length(S)/2));
S1=S1.*hamming(length(S1))';
subplot(2,2,1)
plot(S1)
L=lpc(S1,11);
[H,W]=freqz(1,L);
subplot(2,2,3)
plot(abs(H))
H=abs(H)/max(abs(H));
R=roots(L);
subplot(2,2,4)
zplane(R)
P=[];
for i=1:1:11
    P=[P phase(R(i))]
end
[T]=find(P>=0);
POSP=P(T);
th1=round((1024*80)/FS);
th2=round((1024*100)/FS);
th3=round((1024*150)/FS);
F1=Forcollect(H,W,POSP,th1,FS);
F2=Forcollect(H,W,POSP,th2,FS);
F3=Forcollect(H,W,POSP,th3,FS);
F=[F1 F2 F3];
F=remdup(F);

function [F]=Forcollect(H,W,POSP,th1,FS)
%Forcollect.m
%Formant frequency collection for the particular formant range.
COLH=[]
COLW=[];
j=1;
for i=1:1:length(POSP)
[I,J]= find((W-POSP(i))>0)
```

```
if(I>th1)
COLH{j}=H(I(1)-2*th1:1:I(1)+2*th1);
j=j+1;
COLW=[COLW W(I(1)-1)];
end
end
R=[];
F=[];
for i=1:1:length(COLH)
[M,N]=max(COLH{i});
if ((N>th1)&(COLH{i}(N)>0.1))
%R=[R ((COLH{i}(N)/COLH{i}(N-th1))+(COLH{i}(N)-COLH{i}(N+th1)))/2];
temp=((COLH{i}(N)/COLH{i}(N-th1))+(COLH{i}(N)/COLH{i}(N+th1)))/2 ;
if(temp<1.414)
    F=[F (COLW(i)*FS)/(2*pi)];
else
    F=[F 0];
end
else
if(N>1)
 temp=((COLH{i}(N)/COLH{i}(N-1))+(COLH{i}(N)/COLH{i}(N+1)))/2 ;
if(temp<(1.414/th1))
    F=[F (COLW(i)*FS)/(2*pi)];
else
    F=[F 0];
end
else
    F=[F 0];
end
end
end
```

```
%remdup.m
function [G]=remdup(F)
%To remove duplicate in the array
F=sort(F);
G=[]
for i=1:1:length(F)-1
    if((F(i+1)-F(i))~=0)
       G=[G F(i)];
    end
end
G=[G F(length(F))];
G=nonzeros(G);
```

3.13.2 Formant Extraction Using Homomorphic Filtering

In formant extraction using LPC modeling, the roots of the polynomial $A(Z)$ have to be computed. This increases the computational complexity. The formant extraction using homomorphic filtering has lesser computational complexity compared with LPC modeling technique. The speech signal in Z-domain be represented as $S(Z)$. This is obtained as the product of the excitation source $E(Z)$ and the transfer function of the vocal tract $V(Z)$ (i.e., $S(Z) = E(Z)V(Z)$). For voiced speech segment, $E(Z)$ is assumed to be the periodic impulse stream. The product of $E(Z)$ and $V(Z)$ is converted into additive quantity using log operator as follows.

$$C(Z) = 20 \log_e S(Z) = 20 \log_e E(Z) + 20 \log_e V(Z) \tag{3.26}$$

In $C(Z)$, $20 \log_e E(Z)$ corresponds to the high-frequency component. This is removed by smoothing $C(Z)$ to obtain $C_s(Z) \approx 20 \log_e V(Z)$ term. $V(Z)$ is further obtained as $e^{\frac{C_s(Z)}{20}}$. The peaks of $V(Z)$ around the formant frequency ranges which does not exceed the corresponding formant's bandwidth are declared as the formant frequencies. Figure 3.34 demonstrates the technique of extracting the formant frequencies using homomorphic filtering.

```
%forexthom.m
function [F]=forexthom(S1,FS)
S1=S1.*hamming(length(S1))';
[H,W]=freqz(1,lpc(S1,11),'whole');
figure
```

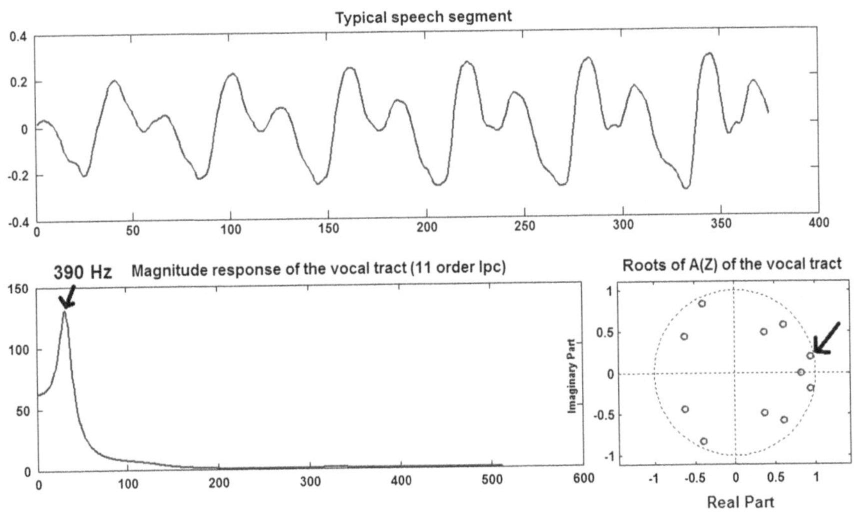

Fig. 3.33 Illustration of the estimation of the formant frequencies using LPC modeling

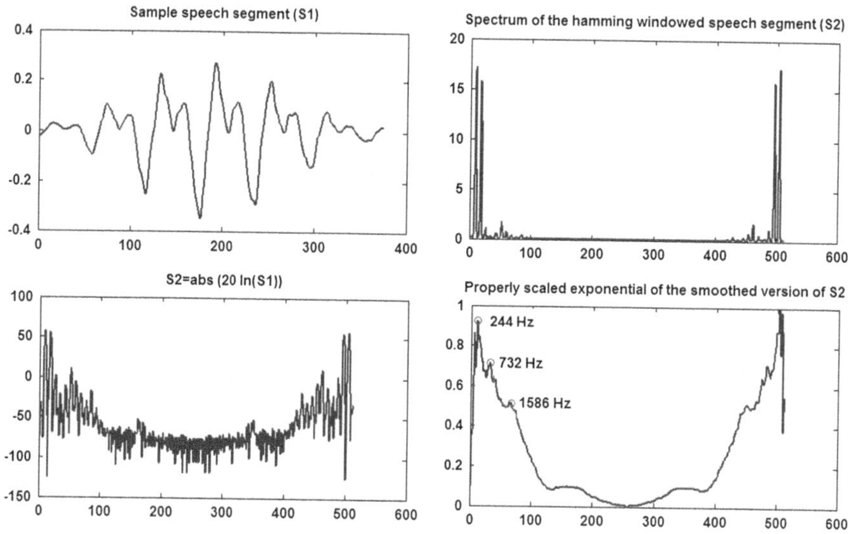

Fig. 3.34 Formant extraction using homomorphic filtering

```
plot(abs(H))
figure
subplot(2,2,1)
plot(S1)
subplot(2,2,2)
plot(abs(fft(S1,512)))
S2=20*log(abs(fft(S1,512)));
subplot(2,2,3)
plot(S2)
S3=smooth(S2,50);
S3=S3+abs(min(S3));
S3=S3/max(S3);
S3=exp(S3)-1;
S3=S3/max(S3);
subplot(2,2,4)
plot(S3)
hold on
th1=round((512*80)/FS);
p=[270/FS 730/FS ]*512;
F1=collectf(p,S3,FS,th1);
 th2=round((512*100)/FS);
 p=[840/FS 2290/FS ]*512;
 F2=collectf(p,S3,FS,th2);
 th3=round((512*150)/FS);
 p=[1690/FS 3010/FS]*512;
```

```
 F3=collectf(p,S3,FS,th1);
 F=[F1 F2 F3];
 F=remdup(F);

%collectf.m
function [formant1]=collectf(p,S3,FS,th1)
cons=th1;
for i=1:1:2
p1=round(p);
while((p1(i)-cons)<=0)
cons=cons-1;
end
[p2,q2]=max(S3(p1(i)-cons:1:p1(i)+cons));
pos1=p1(i)-cons+q2-1;
colpos(i)=pos1;
format(i)=((pos1-1)*FS)/512;
end

[m,n]=max(S3(colpos(1:2)));
n1=n+colpos(1);
if ((n1-th1)<=0)
bwth=(S3(n1)/S3(n1-1)+S3(n1)/S3(n1+1) )/2;
if(bwth<(1.414/th1))
formant1=((colpos(n)-1)*FS)/512;
plot(colpos(n),S3(colpos(n)),'ro')
else
disp('there is no formant ')
end
else
 bwth=(S3(n1)/S3(n1-th1)+S3(n1)/S3(n1+th1) )/2;
 if(bwth<1.4141)
    formant1=((colpos(n)-1)*FS)/512;
plot(colpos(n),S3(colpos(n)),'ro')
else
disp('there is no formant ')
end
end
```

Chapter 4
Speech Compression

Abstract The speech signal is usually sampled with the sampling frequency of 8,000 Hz. If the uniform quantization of 8 bits/sample is used, 64,000 bits are required for 1 s speech data for the sampling frequency of 8,000 Hz. The redundancy in the speech signal is exploited to achieve to the lowest of 3,000 bits for 1 s data. This is known as digital Speech compression. The quality of the speech signal comes down by doing compression. The various techniques like nonuniform quantization, adaptive differential pulse code modulation, code exited linear prediction etc., to compress the speech data are discussed in this chapter. Also the methodology to measure the quality of the compressed speech signal is also discussed in this chapter.

4.1 Uniform Quantization

Let the amplitude of the speech signal is in the range $\frac{-A}{2}$ to $\frac{A}{2}$. If the range is divided into finite number of levels (say N), the quantization step is given as $\delta = \frac{A}{N}$. If the actual sample value is in between $n\delta$ to $(n+\frac{1}{2})\delta$, the sample value is quantized to $n\delta$. Similarly, if the actual sample value is in between $(n+\frac{1}{2})\delta$ to $(n+1)\delta$, the sample value is quantized to $(n+1)\delta$. In either cases, the quantization error is in between $\frac{-\delta}{2}$ to $\frac{\delta}{2}$ and let it be uniformly distributed. The average quantization noise power introduced due to quantization is computed as follows.

$$E(e^2) = \int_{\frac{-\delta}{2}}^{\frac{\delta}{2}} e^2 f(e)\mathrm{d}e = \frac{1}{\delta}\int_{\frac{-\delta}{2}}^{\frac{\delta}{2}} e^2\mathrm{d}e = \frac{\delta^2}{12} \tag{4.1}$$

The number of bits used for the speech signal with the uniform quantization of step size δ is given as $\log_2(N)$.

E. S. Gopi, *Digital Speech Processing Using Matlab*,
Signals and Communication Technology, DOI: 10.1007/978-81-322-1677-3_4,
© Springer India 2014

4.2 Nonuniform Quantization

Let the amplitude of the speech signal ranges from $\frac{-A}{2}$ to $\frac{A}{2}$. The nonuniform quantization is performed as described below:

1. If the actual value is ranging from $\frac{-A}{2}$ to p_1, assign the sample value as q_1.
2. If the actual value is ranging from p_1 to p_2, assign the sample value as them to q_2.
3. In general, if the actual value is ranging from p_i to p_{i+1}, assign the sample value as q_{i+1}.
4. If the actual value is ranging from p_{n-1} to $p_n = \frac{A}{2}$, assign the sample value as q_n.
5. Note that the values of p_i are chosen such that the average error is zero for every interval. (i.e, $p_i = \frac{q_i+q_{i+1}}{2}$).

The optimal values for $q_1 \ldots q_n$ are selected such that average squared quantization noise is reduced as described below. The quantization error is obtained as $\delta(x) = (q_1 - x)$ if x ranges from $\frac{-A}{2}$ to p_1 , $\delta(x) = (q_2 - x)$ if x ranges from p_1 to p_2 and so on. Hence the quantization error δ is treated as the function of x. The average squared quantization noise is computed as follows, where $p(x)$ is the probability density function of the speech signal. Usually, probability density function of the speech signal is modeled as gaussian.

$$E(\delta(x)^2) = \int_{\frac{-A}{2}}^{\frac{A}{2}} \delta(x)^2 p(x) \mathrm{d}x \tag{4.2}$$

$$= \sum_{i=1}^{i=n} \int_{p_{i-1}}^{p_i} p(x)(q_i - x)^2 \tag{4.3}$$

To obtain the optimal values of q_i that minimize $E(\delta(x)^2)$, differentiate (4.3) with respect to q_m for arbitrary m and equate to zero.

$$2 \int_{p_{i-1}}^{p_i} p(x)(q_i - x) \mathrm{d}x = 0 \tag{4.4}$$

$$q_i \int_{p_{i-1}}^{p_i} p(x) \mathrm{d}x = \int_{p_{i-1}}^{p_i} p(x) x \mathrm{d}x \tag{4.5}$$

$$\Rightarrow q_i = \frac{\int_{p_{i-1}}^{p_i} p(x) x \mathrm{d}x}{\int_{p_{i-1}}^{p_i} p(x) \mathrm{d}x} \tag{4.6}$$

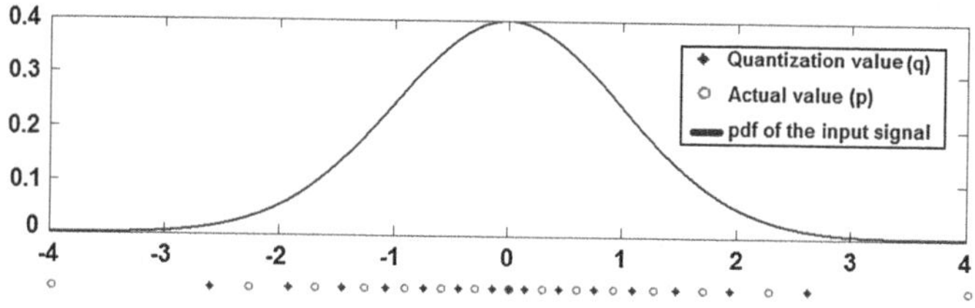

Fig. 4.1 Nonuniform quantization points obtained for the signal that is gaussian distributed with $mean = 0$ and $variance = 1$

Table 4.1 Nonuniform quantization for gaussian distributed signal with $mean = 0$ and $variance = 1$

Actual range	Quantization value	Code word number
−4.0000 to −2.2700	−2.6100	−7
−2.2700 to −1.6957	−1.9300	−6
−1.6957 to −1.2714	−1.4615	−5
−1.2714 to −0.9144	−1.0814	−4
−0.9144 to −0.5934	−0.7474	−3
−0.5934 to −0.2921	−0.4393	−2
−0.2921 to 0	−0.1450	−1
0	0	0
0 to 0.2921	0.1450	1
0.2921 to 0.5934	0.4393	2
0.5934 to 0.9144	0.7474	3
0.9144 to 1.2714	1.0814	4
1.2714 to 1.6957	1.4615	5
1.6957 to 2.2700	1.9300	6
2.2700 to 4.0000	2.6100	7

Thus, the optimal values of q_i are obtained as follows:

1. Initialize the values for p_i for $i = 0$ to $i = n$. Note that $p_0 = \frac{-A}{2}$ and $p_{nf} = \frac{A}{2}$.
2. Compute q_i using (4.6).
3. Compute $p_i = \frac{q_i + q_{i+1}}{2}$ for $i = 1$ to $i = n - 1$.
4. Repeat steps (2) and (3) until the value converges.

The nonuniform quantization obtained for gaussian distributed signal with $mean = 0$ and $variance = 1$ is illustrated in the Fig. 4.1. The quantization points are referred as the code words ranging from −7 to 7. The quantization values and the corresponding code words are tabulated as Table 4.1. Note that symmetry is maintained with respect to orgin.

```
%nuqg.m
%15 point nonuniform quantization achieved for
%gaussian distributed data with unit variance.
%Initializing the points between -4 to 4
p=rand(1,6)*4;
```

```
p=sort(p);
p=[0 p 4];
%Computation of q's
for iteration=1:1:100000
for i=1:1:7
    t=p(i):0.001:p(i+1);
    temp=(1/sqrt(2*pi))*exp(-(t.^2)/2);
    q(i)=(eps+trapz(t.*temp))/(eps+trapz(temp));
end
p(1)=0;
p(8)=4;
for i=2:1:7
    p(i)=(q(i-1)+q(i))/2;
end
end
figure
subplot(2,1,1)
t=-4:0.001:4;
a=(1/sqrt(2*pi))*exp(-(t.^2)/2);
plot(t,a)
subplot(2,1,2)
p=p(2:1:length(p));
p=[-p 0 p];
p=sort(p);
q=[-q 0 q];
q=sort(q);
plot(p,zeros(length(p)),'ro')
hold on,plot(q,zeros(length(q)),'b*')
```

Most commonly used nonuniform quantizations are μ-law and A-law. The nonuniform quantization is achieved by passing the speech signal through the compressor and applying the uniform quantization. Let the speech signal s_i amplitude is normalized over the range -1 to 1. The normalized output of the compressor represented as s_o is related to s_i for μ-law and A-law and are given in (4.7) and (4.8–4.9), respectively (refer Fig. 4.2)

$$|s_o| = \frac{log(1+\mu|s_i|)}{1+\mu} \tag{4.7}$$

$$|s_o| = \frac{A|s_i|}{1+log(A)}, 0 \le |s_i| \le \frac{1}{A} \tag{4.8}$$

$$|s_o| = (1+logA)|s_i|, \frac{1}{A} \le |s_i| \le 1 \tag{4.9}$$

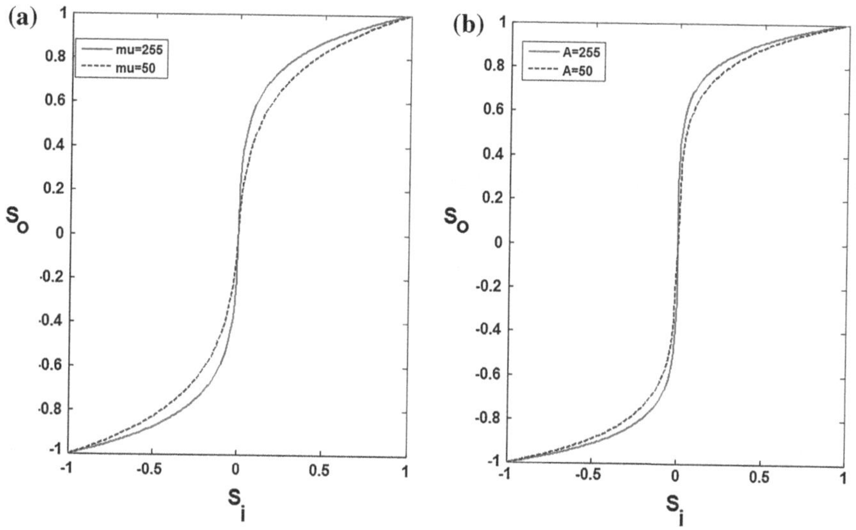

Fig. 4.2 Illustrating the relation between input and output for **a** μ-law companding **b** A-law companding

4.3 Adaptive Quantization

In the nonuniform quantization, it is seen that the quantization step is less when the amplitude of the speech signal is lower and vice versa. Instead of fixing up the step size (as the function of the amplitude of the speech signal) like A-law and μ-law, the step size can be adaptively chosen in the real time as described below.

1. Initialize the value of the scaling factor as $k = 1$. The first sample value of the speech signal is assumed to be obtained from the signal that is gaussian distributed with $mean = 0$ and $variance = 1$. Quantize the first value with the nonuniform quantization mentioned in the Table 4.1.
2. If the codebook used to quantize the first sample is 1, k is updated as $k = m_1 \times k$. Divide the second sample with k and code the obtained value using the nonuniform quantization mentioned in the Table 4.1. If the codebook used to quantize the first sample is 7, k is updated as $k = m_7 \times k$. Divide the second sample with k and code the obtained value using the nonuniform quantization mentioned in the Table 4.1. In general, k is updated based on the codebook used to quantize the first sample. Divide the second sample with k and code the obtained value using the nonuniform quantization mentioned in the Table 4.1.
3. Based on the codebook used to quantize the second sample, the scaling factor k is updated. Divide the third sample with k and code the obtained value using the nonuniform quantization mentioned in the Table 4.1.
4. In general based on the codebook used to quantize the previous sample, k is updated. Divide the next sample with k and code the obtained value using the nonuniform quantization mentioned in the Table 4.1.

5. The multiplication factors (m_i) to update k for the various code book numbers are chosen such that the scaled input sample fits to the gaussian distributed ($mean = 0$ and $variance = 1$) signal.

4.4 Differential Pulse Code Modulation

The individual frame of the speech signal (speech segment) is highly correlated. Hence, the nth sample of speech segment can be predicted with the previous N samples with minimum error. Let the nth sample of the speech signal be represented as $s(n)$. It is predicted using the previous N samples as follows: $\widehat{s}(n) = a_1 s_q(n - 1) + a_2 s_q(n - 2) + a_3 s_q(n - 3) + \cdots + a_N s_q(n - N)$, where $a_1, a_2, \ldots a_N$ are the prediction coefficients. Also note that the prediction is obtained using the quantized previous samples represented as $s_q(n-1), s_q(n-2), \ldots s_q(n-N)$. The prediction error represented as $e(n) = s(n) - \widehat{s}(n)$ is quantized to obtain $e_q(n)$ and it is transmitted. In the receiver section, the prediction of $x(n)$ is done using the prediction coefficients and the quantized error $e_q(n)$ is added with $\widehat{s}(n)$ to obtain the quantized sample $s(n)$ which is represented as $s_q(n)$.

4.4.1 *Illustrations of the Prediction of Speech Signal Using lpc*

The speech signal is divided into frames of length 20 ms. Four frames are treated as one speech segment. For every speech segment, 11th order lpc are obtained using the first frame. Using the coefficients, speech samples are predicted for the corresponding speech segment. Illustration of prediction of the speech samples using lpc is given in the Fig. 4.3 and the zoomed version is given in the Fig. 4.4. Figure 4.5 shows the error in prediction and the corresponding histogram. It is observed that histogram follows the gaussian distribution plot.

```
%prediction.m
function [res]=prediction(x,framesize)
[coef]=lpc(x(1:1:framesize),11);
coef=coef(2:1:11);
xp=0;
for n=12:1:length(x)
    temp=0;
     for j=1:1:10
     temp=temp-coef(j)*x(n-j);
      end
      xp(n)=temp;
end
res1=xp(12:1:length(xp))';
res2=x(12:1:length(x))';
 res=[res1 res2];
```

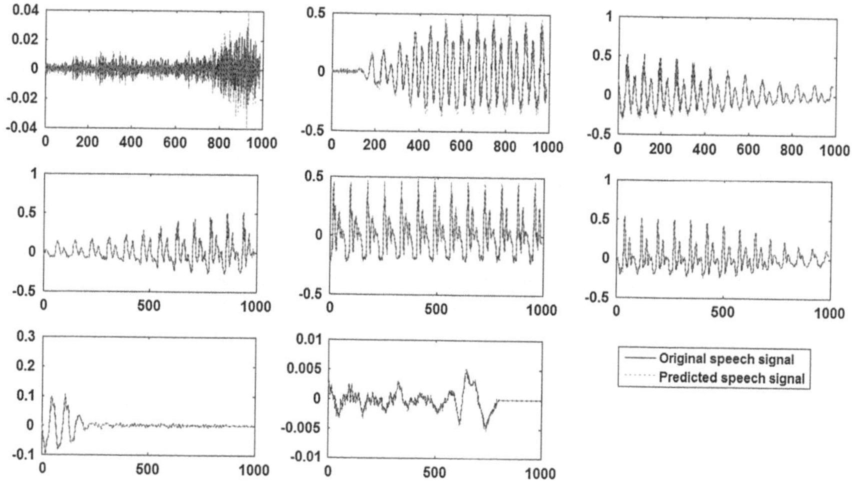

Fig. 4.3 Illustration of prediction of the speech samples for 8 speech segments

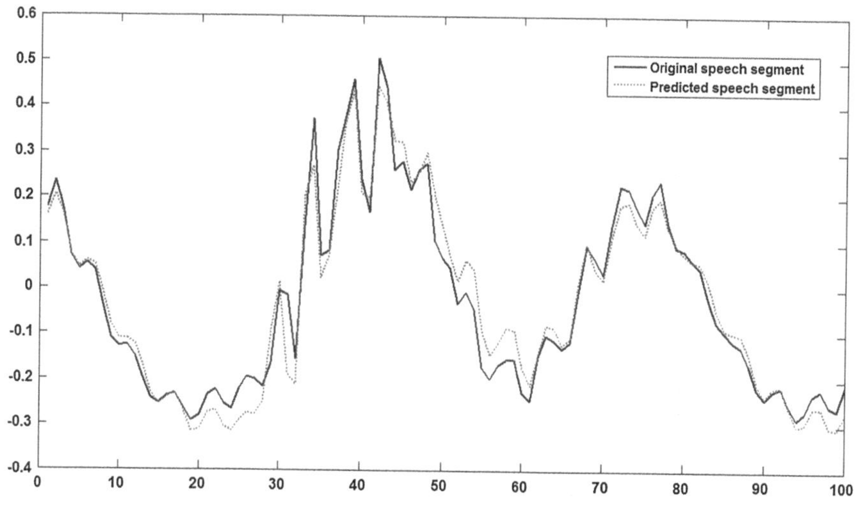

Fig. 4.4 Zoomed version to illustrate the prediction of 100 speech samples of the 3rd segment

The steps involved in DPCM for every speech segment (s) with length L (say) are summarized below:

1. Compute the kth order lpc coefficients using the first frame of the speech segment.
2. Quantize $s(n)$ as $s_q(n)\ \forall\ n = 1 \ldots k$.
3. Compute $\widehat{s}(k+1) = a_1 s_q(k) + a_2 s_q(k-1) + a_3 s_q(k-2) + \cdots + a_k s_q(1)$.
4. Compute the error $e(k+1) = s(k+1) - \widehat{s}(k+1)$.
5. Quantize the error $e(k+1)$ as $e_q(k+1)$.
6. Obtain $s_q(k+1) = \widehat{s}(k+1) + e_q(k+1)$.

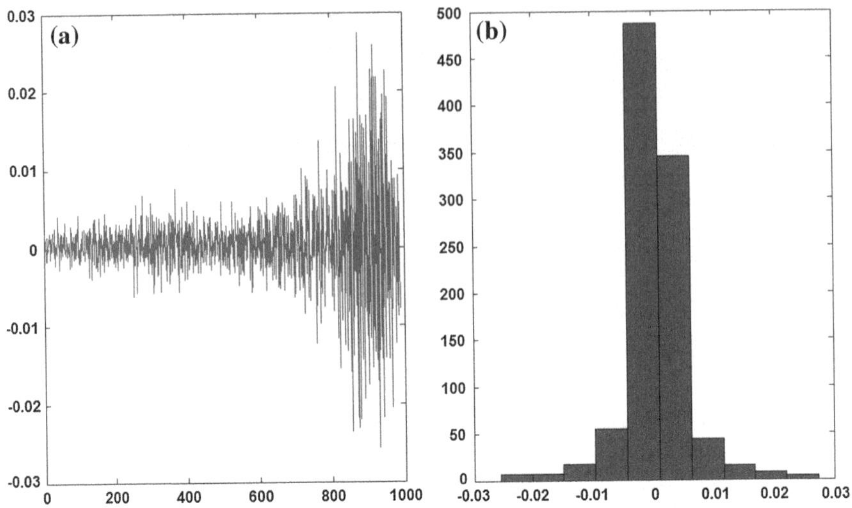

Fig. 4.5 **a** Error plot **b** Corresponding histogram

7. Repeat the steps (3)–(5) to obtain $e_q(m)\forall m = k+2\ldots L$.
8. $s_q(n)\forall n = 1\ldots k$ and $e_q(m)\forall m = k+2\ldots L$ forms the compressed data for the corresponding speech segment.

In DPCM, if on-uniform quantization is applied for the error signal, the technique is known as Adaptive Differential pulse code modulation. Also the filter model used for prediction can be either Auto regressive (AR) filter or Auto-regressive moving average (ARMA).

4.5 Code-Excited Linear Prediction

In Code-Excited Linear Prediction (CELP), the source-filter model is exploited to produce the speech segment. The constants used to produce the particular speech segment is stored and hence highest compression ratio is achieved using CELP. The steps involved in speech compression using CELP are summarized below:

1. The speech signal is divided into frames of 30 ms each. For the sampling frequency of 11, 250, frame length of 375 is used. The vocal tract is modeled using the lpc for every frames. Every frame is multiplied with the hamming window to avoid Gibbs phenomenon. The bandwidth of the vocal tract transfer function $V(Z)$ is increased by the factor 0.994 to avoid chirps due to very sharp formant peaks. The bandwidth is increased by replacing Z with $\frac{Z}{0.994}$. Rewriting the transfer function of the vocal tract by replacing Z with $\frac{Z}{0.994}$ is as follows:

$$V(Z) = \frac{G}{1 - \sum_{k=1}^{k=r} \frac{a_k}{0.994^{-k}} Z^{-k}} \tag{4.10}$$

Thus the lpc a_k are modified as $a_k 0.994^k$ to increase the bandwidth by 0.994.

2. The speech is synthesized for every 7.5 ms ($\frac{1}{4}$ of the frame length). This is known as speech segment. For the sampling frequency of 11, 250, the length of the speech segment is 94 samples.
3. The excitation source is passed through the vocal tract filter to model every speech segment. The excitation source for the particular speech segment is assumed to be consists of summation of the linear combinations of the random noise signal (from the codebook) and the delayed version of the excitation source for the previous speech segment.
4. The codebook consists of 1,082 samples filled up with $+1$, -1, and 0, with the probability of 0.1, 0.1, and 0.8, respectively. The plot of the typical codebook and the corresponding histogram are shown the Fig. 4.6. The content of the codebook is indexed numerically from 1 to 1,082. The random noise signal with index r is referred to the content of the codebook from r to $r + length\ of\ the\ speech\ segment - 1$. The best code index number that minimizes the error in speech synthesis is chosen for every speech segment.
5. Thus the modified lpc (for every frame), codebook index number(I), the value of the delay constant for the previous excitation source (D) (computed for every frame), gain for the random noise signal ($G1$), and gain for the delayed version of the excitation source for the previous speech segment ($G2$) are stored for every segment. Thus for every segment, apart from the lpc, the number of constants used are 3. For every frame, the previous excitation is also stored in terms of codebook index (PI). If the number of lpc used is 10, the total number of constants used to generate one frame of the speech signal is calculated as $10 + 4 \times 3 + 2 = 24$. Apart from this, the common codeword with 1, 082 samples are stored for the

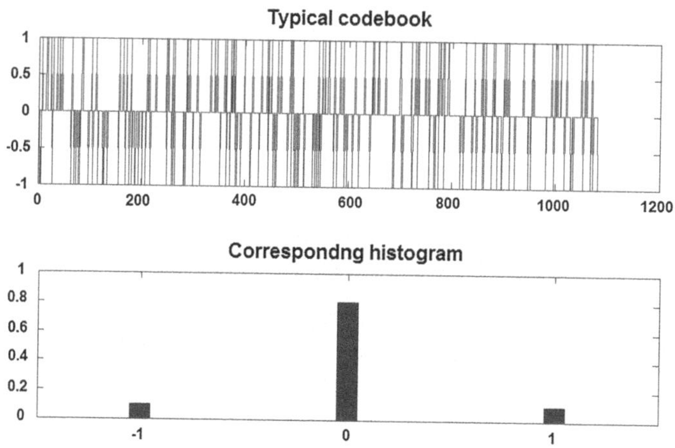

Fig. 4.6 Typical codebook and the corresponding histogram

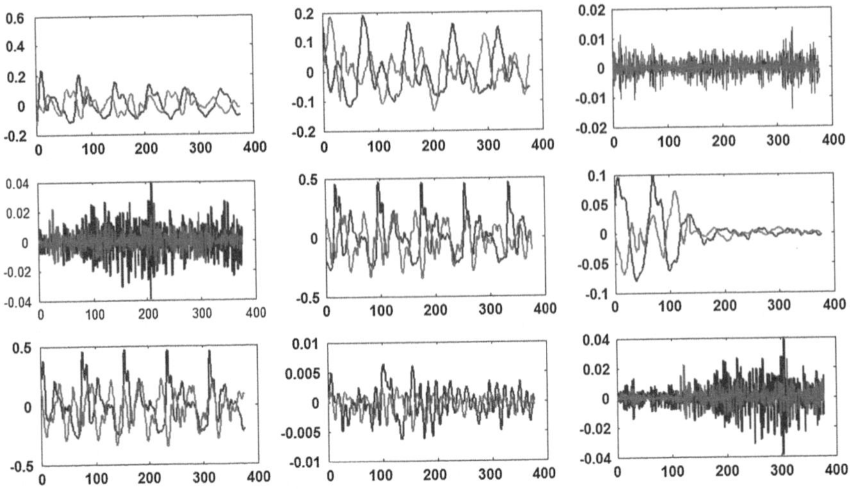

Fig. 4.7 Snapshots of the original speech segments and the corresponding synthesized speech segments

entire speech synthesis. For the 20 frame speech signal, the total number of samples used are 7,500 without CELP. After CELP, the number of constants used are $19 \times 20 = 380$ (codeword length is not used for calculation). Approximately 20 : 1 compression ratio is achieved using CELP.

6. The typical speech signal and the synthesized signal is given in Figs. 4.7, 4.8.

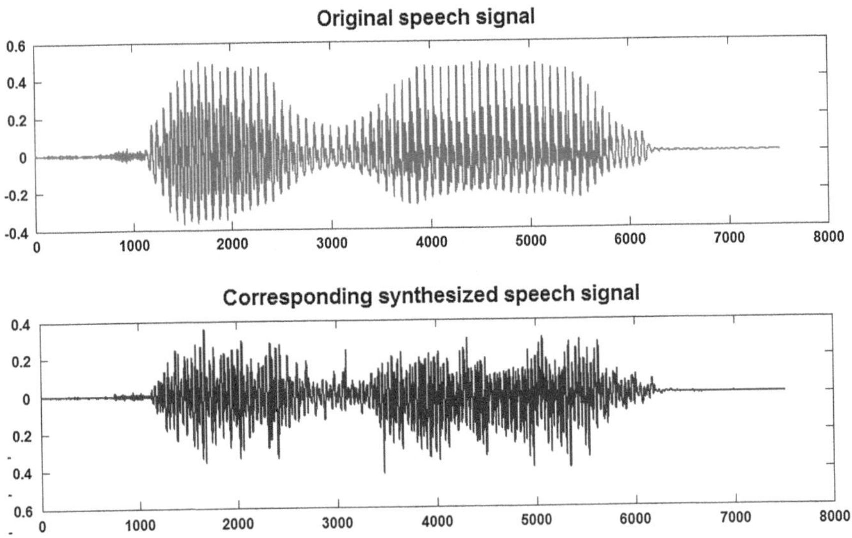

Fig. 4.8 Illustration of the original isolated speech signal (word) and the corresponding synthesized speech segment

4.5.1 Estimation of the Delay Constant D

The source signal (excitation signal) consists of linear combinations of the source collected from the codebook and the delayed version of the previous excitation signal. The delay is constant for every frame. The delay constant for the ith frame of the speech signal is estimated as follows: Let the ith frame for the speech segment be represented as $s_i(n)$ and the corresponding vocal tract transfer function is represented as $V(Z)$ (after bandwidth increment). The source signal (excitation signal) is represented as $E_i(n)$. They are related as $S_i(Z) = E_i(Z)V(Z)$. Let $A(Z) = \frac{1}{V(Z)} = 1 - \sum_{k=1}^{k=r} \frac{a_k}{0.994^{-k}} Z^{-k}$. Hence $E_i(Z) = S_i(Z)A(Z) \Rightarrow e_i(n) = s_i(n) * a(n)$. The periodicity of the exitation signal $E_i(Z)$ is the delay constant for the ith frame. The speech frame, corresponding vocal tract, and the corresponding excitation signal are given in the Fig. 4.9. Given the excitation signal, the periodicity is computed as follows.

1. The frequency of the excitation signal varies from 54 to 400 Hz. For the sampling frequency of 12, 500, it is from 32 to 232 samples.
2. Initialize $k = 32$ and $n = 1$. Collect the entire excitation signal from nth sample.
3. Divide the excitation signal into subblocks with k samples in each subblock. Compute the euclidean distance between the consecutive subblocks. Let it be $\text{error}_{k,1}$.
4. Repeat the third step for $n = 2$ to $n = k - 1$ to obtain $\text{errror}_{k,2}$ to $\text{error}_{k,k-1}$. Identify the $\min_{i=1}^{i=k-1} \text{error}_{k,i}$ and store the value as error_k.
5. Repeat steps 3 and 4 for k varies from 32 to 232.
6. Declare that $\arg(\min_{k=32}^{k=232} \text{error}_k)$ as the delay factor D for the particular speech frame.

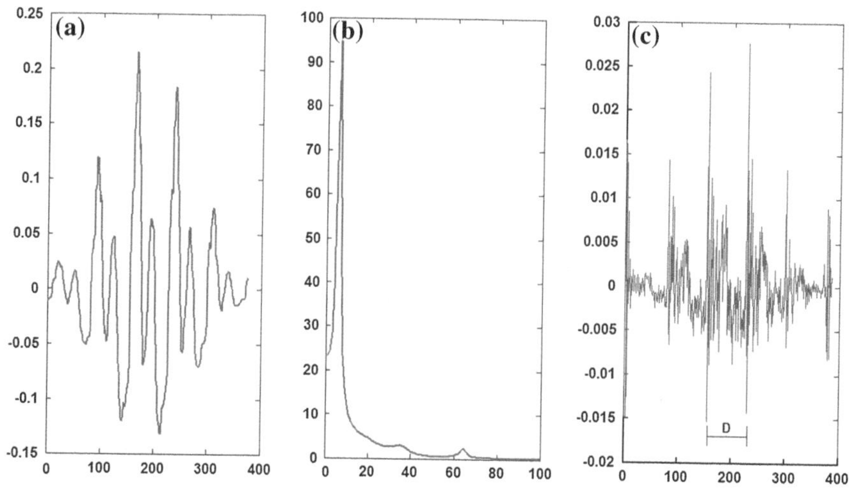

Fig. 4.9 Illustration of the computation of the periodicity D of the excitation signal in CELP technique **a** Speech frame **b** Vocal tract **c** Corresponding excitation signal

4.5.2 Estimation of the Gain Constants $G1$ and $G2$

The gain $G1_i$ is computed such that prediction error of the excitation signal $E_i(n)$ is minimized as follows. Let the prediction error of the ith excitation signal is represented as PE_i and is given as $PE_i(n) = E_i(n) - G1_i E_i(n - D)$. D is estimated as described in the Sect. 4.5.1. The $G1_i$ is estimated such that the average squared prediction error is minimized. $E(PE^2) = E((E_i(n) - G1_i E_i(n - D))^2)$ is minimized, where E is the expectation operator. Differentiating $E(PE^2)$ with respect to $G1_i$ and equate to zero, we get the following.

$$E(E_i(n)^2) = G1_i E(E_i(n)E_i(n - D))$$
$$\Rightarrow G1_i = \frac{E(E_i(n)^2)}{E(E_i(n)E_i n - D)} = \frac{R_E(0)}{R_E(D)}$$

Once $G1_i$ is obtained, the $G2_i$ is estimated as follows. Let the ith current codebook source is represented as C_i and the delayed version of the $(i - 1)$th excitation (total source) is represented as E_{i-1}. The E_i is related to E_{i-1} and C_i as $E_i = G1C_i + G2E_{i-1}$. The synthesized speech signal $SY_i(n)$ is obtained by passing the excitation signal through the filter $V(Z)$. The error in speech production is obtained as $\mathrm{ERROR}_i(Z) = E_i(Z)V(Z) - S_i(Z)$. The Gain constants are estimated such that the weighted error is minimized (i.e., $\mathrm{ERROR}(Z)W(Z)$ is minimized). The $W(Z)$ is chosen as follows (refer Fig. 4.10)

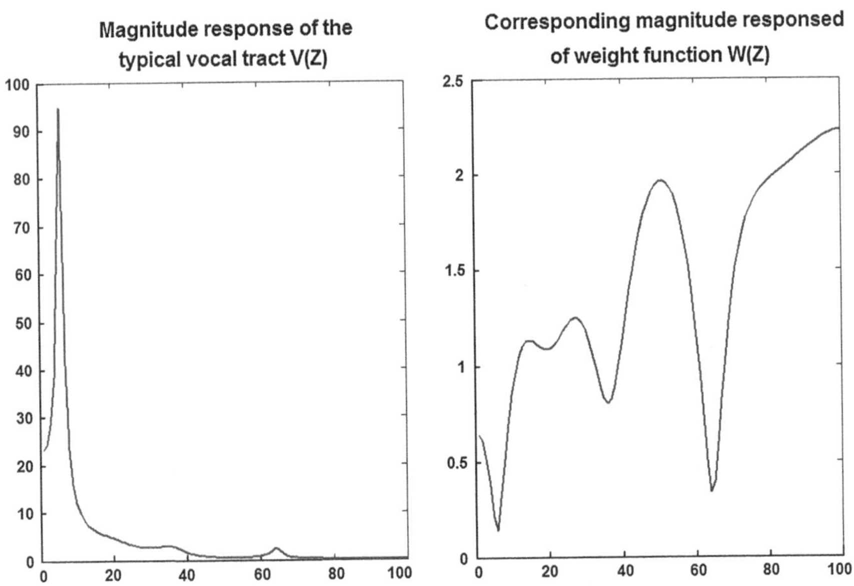

Fig. 4.10 Illustration of the weight function $W(Z)$ used in CELP

$$W(Z) = \frac{V\left(\frac{Z}{0.8}\right)}{V(Z)} \tag{4.11}$$

The error $\mathrm{ERROR}_i(Z)W(Z)$ is minimized as follows:

$$\mathrm{ERROR}_i(Z)W(Z) = (E_i(Z)V(Z) - S_i(Z))W(Z)$$
$$\Rightarrow \mathrm{ERROR}_i(Z)W(Z) = E_i(Z)V(Z)W(Z) - S_i(Z)W(Z)$$
$$\mathrm{ERROR}_i(Z)W(Z) = G1_iC_i(Z)V(Z)W(Z) + G2_iE_{i-1}V(Z)W(Z) - S_i(Z)W(Z)$$

Let $Y(Z) = E_{i-1}V(Z)W(Z)$ and $Q(Z) = S_i(Z)W(Z) - G_1C_i(Z)V(Z)W(Z)$. The weighted error in z-domain and in time domain is represented as $G2_iY(Z) - Q(Z)$ and $G2_iy(n) - q(n)$, respectively. $G2_i$ is optimized by minimizing $E((G2_iy(n) - q(n))^2)$, where E is the expectation operator. Differentiating $E((G2_iy(n) - q(n))^2)$ with respect to $G2_i$ and equate to zero, we get the following:

$$G2_i = \frac{E(y(n)q(n))}{E(y(n)^2)} \tag{4.12}$$

```
%celp.m
function [synsp,codebook,ICN,LPC,CN,G1,G2,DELAY]=celp(S,FS);
%ORIS-Original speech signal
%synsp-Synthesized speech signal
%ICN-Initial code number
framelength= round(FS*30*10^(-3));
excitationlength=round(FS*7.5*10^(-3));
%Generating the code book
r=rand(1,1082);
codebook=[];
for i=1:1:1082
if(r(i)<=0.1)
    codebook=[codebook -1];
elseif ((0.1<r(i))&(r(i)<0.9))
    codebook=[codebook 0];
else
    codebook=[codebook 1];
end
end
CN=[];
ICN=[];
G1=[];
G2=[];
DELAY=[];
synsp=[];
ORIS=[];
for frame=1:1:fix(length(S)/framelength)
```

```
S1=S((frame-1)*framelength+1:1:(frame-1)*framelength+framelength+1) ;
l=lpc(S1,10);
LPC{frame}=l;
l1=l(2:1:length(l));
l1=l1.*(0.994.^[1:1:(length(l)-1)]);
lmod1=[1 l1];
l2=l(2:1:length(l));
l2=l2.*(0.8.^[1:1:(length(l)-1)]);
lmod2=[1 l2];
%Usually frequency of the excitation is in the range of 54 Hz to 400 Hz.
%Calculated in terms of number of samples as the range from 32 to 232
ee=conv(S1,l);
col1=[];
for n=32:1:232
    col2=[];
for k=1:1:n
temp=ee(k:1:length(ee));
m=fix(length(temp)/n);
if(m==1)
    break
end
f=temp(1:1:m*n);
r=reshape(f,n,m);
error=0;
for p=1:1:size(r,2)-1
error=error+sum((r(:,p)-r(:,p+1)).^2)/((m-1)*n);
end
col2=[col2 error];
end
[s,t]=min(col2);
col1=[col1 s];
end
[u1,v1]=min(col1);
m=v1+31;
DELAY=[DELAY m];
%The excitation period is m/FS or excitation frequency is FS/m
r=rand;
ICN=[ICN round(r*(1082-excitationlength))+1];
e=codebook(round(r*(1082-excitationlength))+...
1:1:round(r*(1082-excitationlength))+excitationlength);
for excitation=1:1:4
%gain1=Rsm/Rs0
 Rsm=sum([e zeros(1,m)].*[zeros(1,m) e]);
 Rsm=Rsm/length([e zeros(1,m)]);
 Rs0=sum(e.*e);
 Rs0=Rs0/length(e);
 gain1=(Rsm+eps)/Rs0;
 G1=[G1 gain1];
 %gain2  is computed
```

```
  t=filter(1,lmod2,S1((excitation-1)*excitationlength+1:1:(excitation-1)*...
  excitationlength+excitationlength));
  q=filter(1,lmod2,gain1*[zeros(1,m) e]);
 L1= length(q);
 L2=length(t);
 if(L1>=L2)
    u=t-q(1:1:length(t));
 else
   u= t(1:1:length(q))-q;
 end
 error1=[];
 for iteration=1:1:length(codebook)-excitationlength-1;
 codeselection=codebook(iteration:1:iteration+excitationlength-1);
 y=filter(1,lmod2,codeselection);
 u=u(1:1:length(u));
 gain2=sum(y.*u)/sum(y.^2);
 G2=[G2 gain2];
 temp1=[zeros(1,m) e];
 temp1=temp1(1:1:excitationlength);
 temp2=gain1*temp1+gain2*codeselection;
 sp=filter(1,1,temp2);
 z1=S1((excitation-1)*excexcitationitationlength+1:1:(excitation-1)*...
 excitationlength+excitationlength)-sp(1:1:excitationlength);
 e2=filter(1,lmod2,z1);
 error1=[error1 sum(e2.^2)];
 end
 [a,b]=min(error1);
 selectedcode=codebook(b:1:b+excitationlength-1);
 y=filter(1,lmod2,selectedcode);
 u=u(1:1:length(u));
 gain2=sum(y.*u)/sum(y.^2);
 temp1=[ zeros(1,m)  e];
 temp1=temp1(1:1:excitationlength);
 temp2=gain1*temp1+gain2*codeselection;
 sp=filter(1,1,temp2);
 sp=sp(1:1:excitationlength);
 synsp=[synsp sp];
 ORIS=[ORIS S1((excitation-1)*excitationlength+1:1:(excitation-1)*...
 excitationlength+excitationlength)];
 e= temp2;
 end
 end
 figure
 subplot(2,1,1)
 plot(ORIS)
 hold on
 subplot(2,1,2)
 plot(synsp,'r')
 j=1;
```

Table 4.2 List of various speech compression techniques

Type	Year	MOS	Bits per 8000 samples (approx) (K)
Pulse Code Modulation (PCM16)	–	5	128
A-law and μ-law	1972	5	64
Adaptive Differential Pulse Code Modulation (ADPCM)	1984	5	32
Linear Predictive Co-efficients (LPC10)	1984	2	2
Global System for Mobile communication (GSM)	1987	4	16
Code Exited Linear Prediction (CELP)	1991	3	8
Global System for Mobile communication (GSM2)	1994	4	8
Low Delay CELP (LD-CELP)	1994	5	16
CELP	1996	5	8

```
for i=1:1:9
r=round(rand*(length(synsp)-4*excitationlength-1));
subplot(3,3,j)
plot(ORIS(r:1:r+4*excitationlength))
hold on
plot(synsp(r:1:r+4*excitationlength),'r-')
j=j+1;
end
```

4.6 Assessment of the Quality of the Compressed Speech Signal

The quality of the compressed speech signal is measured by the subjective measurement (rate) given by the group of test listeners. The mean of the collected data is the measure of the quality of the compressed speech signal. This is known as Mean Opinion Score (MOS) For instance, MOS with value 1 indicates poor, 2 indicates fair, 3 indicates average, 4 indicates good, and 5 indicates excellent. The summary of various speech compression techniques applied is tabulated in Table 4.2

Appendix A
Constrained Optimization Using Lagrangian Techniques

A.1 Constrained Optimization with Equality Constraints

Identifying the unknown variables that maximizes or minimizes the function f subject to the constraints $g = 0$ is solved using Lagrangian technique. The optimal solution is obtained as the extremal point where $\Delta f + \lambda \Delta g = 0$. This is explained using the following example.

A.1.1 Example 1

The problem is to minimize the function $f(x, y) = 2x^2 + 4y + 3$ subject to the constraint $g(x, y) = x + y + 3 = 0$. The gradient $\nabla f = [4x\ 4]^T$ and $\nabla g = [1\ 1]^T$. At the optimal point, we obtain the following:

$$\nabla f + \lambda \nabla g = [4x\ 4]^T + \lambda [1\ 1]^T = 0 \tag{A.1}$$

Solving we obtain $\lambda = -4$ and $x = 1$. Thus, the optimal solution is $(1, -4)$ and the corresponding value of the function $f(x, y)$ is -11. This is illustrated in Fig. A.1. The contours are drawn for various levels for the function f. The optimal point is on the black line. Red-colored lines are the gradient of the function f and g. It is seen that they are parallel at $x = 1$. Also, it is seen that the value of the function f is increasing on either side of the point $(1, -4)$ on the line g. The zoomed portion to illustrate the solution is shown in Fig. A.2. Hence, the obtained extremal point is the minima point.

```
%larangeandemo.m
%demonstration for equality constraints
i=1;
j=1;
for x=-5:1:5
```

E. S. Gopi, *Digital Speech Processing Using Matlab*,
Signals and Communication Technology, DOI: 10.1007/978-81-322-1677-3,
© Springer India 2014

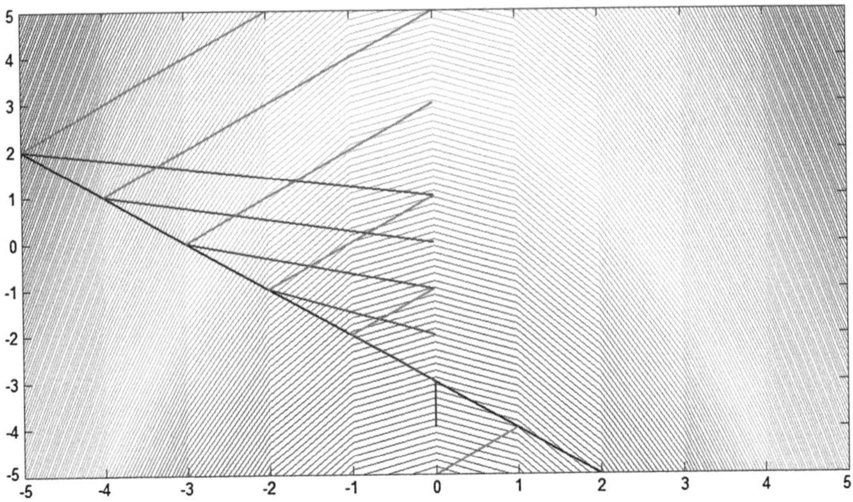

Fig. A.1 Illustration of the constrained optimization using Lagrangian technique. Note that the *X* and *Y* axis are not having identical scale

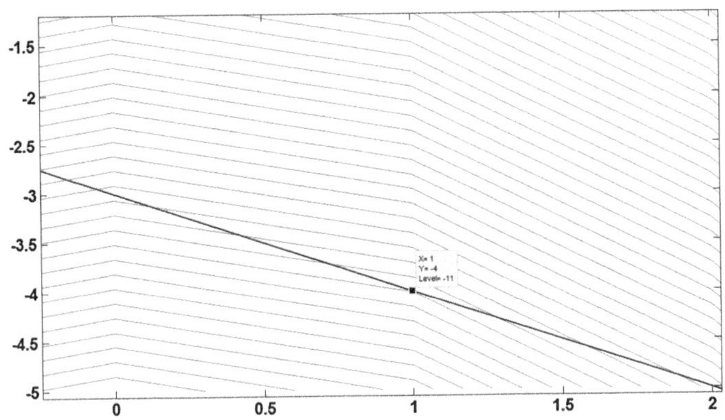

Fig. A.2 Zoomed portion of Fig. A.1

```
    i=1;
for y=-5:1:5
z(i,j)=2*x^2+4*y+3;
i=i+1;
end
j=j+1;
end
 x=-5:1:5;
 y=-5:1:5;
contour(x,y,z,150)
```

```
x=-5:1:5;
y=-3-x;
hold on
plot(x,y,'k')
for x=-5:1:5
plot([x 0],[ (-3-x) (-4-x) ],'r')
plot([x 0],[(-3-x) (-3-2*x)],'m')
hold on
end
```

A.2 Constrained Optimization with Inequality Constraints

Consider the problem of obtaining the optimal point that maximizing the function $f(x, y) = 2x^2+4y+3$, such that it satisfies the condition $g(x, y) = x+y^2-2 > 0$. The extremal points are obtained as follows $\nabla f + \lambda\nabla = 0$. Solving we get the following:

$$\nabla f + \lambda \nabla g = [4x\ 4] + \lambda[1\ 2y] = 0 \tag{A.2}$$

$$\Rightarrow \lambda^3 + \lambda^2 - 16 = 0 \Rightarrow \lambda = -1.5784 \tag{A.3}$$

$$(or)\lambda = -7.7324\ (or)\ \lambda = 1.3109 \tag{A.4}$$

In the subplots (a) and (b) of Fig. A.3, pink color indicates the gradient vector of the function $f(x, y)$ and the red color indicates the gradient vector of the function $g(x, y)$ for every point of x. Also note that the gradient vectors are computed with the positive square root for solving y. In (a), the gradients are parallel at $x = 0.4$ and $y = 1.3$. This corresponds to $\lambda = -1.5784$ and the functional value $f = 8.52$. In (b), the gradients are parallel at $x = 1.9$ and $y = 0.3162$. This corresponds to $\lambda = -7.7324$ and the functional value $f = 11.42$. Maximum value of f are obtained at the top in the contour diagram. But we need to select the point in the region as shown in Fig. A.3d. This implies that the optimum value corresponds to $\lambda = -7.7324$. In subplot (c), the gradient vectors are computed with the negative square root for solving y. In (c), the gradients are parallel at -0.3 and -1.5. This corresponds to $\lambda = 1.3109$. In this case, the functional value obtained as $f = -2.82$. This corresponds to the minimal value in the region as shown in Fig. A.3. In general, we have to choose positive lambda to obtain minima point and the negative lambda to obtain maxima point. This is Kuhn–Tucker condition, which is summarized below.

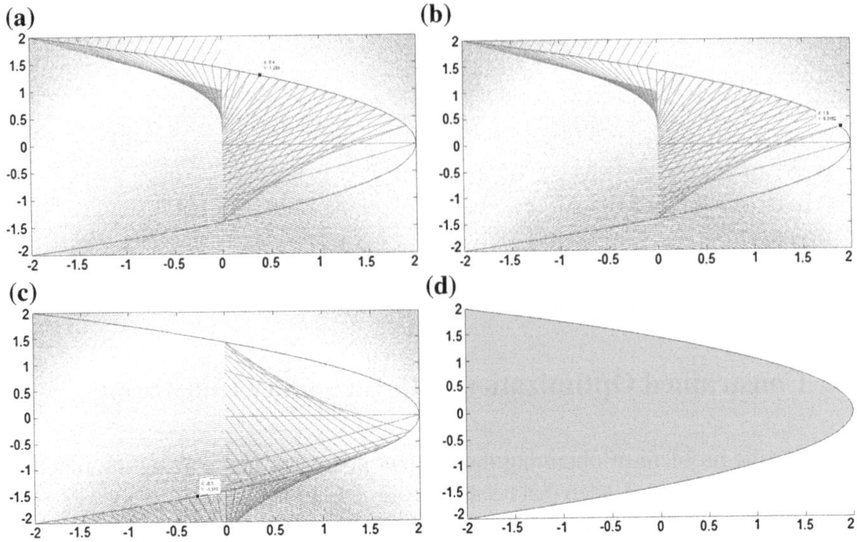

Fig. A.3 Illustration for inequality constraint. **a** Optimum point obtained for $\lambda = -7.7324$, **b** optimum point obtained for $\lambda = -1.5784$, **c** optimum point obtained for $\lambda = 1.3109$, **d** illustrating the region $x + y^2 - 2 > 0$

A.3 Kuhn–Tucker Conditions

Consider the constrained optimization problem as given below. Minimize the function $f(x_1, x_2, \cdots, x_n)$, subject to the constraints $g_1(x_1, x_2, \cdots, x_n) = 0$ and $g_2(x_1, x_2, \cdots, x_n) \geq 0$ and $g_3(x_1, x_2, \cdots, x_n) \leq 0$. The Lagrangian equation is formulated as follows. $f(x_1, x_2, \cdots, x_n) + \lambda_1 g_1(x_1, x_2, \cdots, x_n) + \lambda_2 g_2(x_1, x_2, \cdots, x_n)$ $-\lambda_3 g_3(x_1, x_2, \cdots, x_n) = 0$. Differentiating the above equation with respect to $x_1, x_2, \cdots, x_n$ and equate to zero to obtain n equations. Also, the Lagrangian multiplier satisfies the following conditions: $\lambda_2 \geq 0$, $\lambda_3 \geq 0$ and $\lambda_2 g_2(x_1, x_2, \cdots, x_n) + \lambda_3 g_3(x_1, x_2, \cdots, x_n) = 0$. Along with this, $g_1(x_1, x_2, \cdots, x_n) = 0$ is used to obtain the extremal points.

```
%kuhntucker.m
%demonstration for inequality constraints
figure
i=1;
j=1;
for x=-2:0.1:2
    i=1;
for y=-2:0.1:2
z(i,j)=2*x^2+4*y+3;
i=i+1;
end
```

```
j=j+1;
end
 x=-2:0.1:2;
 y=-2:0.1:2;
contour(x,y,z,150)
y=-2:0.1:2;
x=2-y.^2;
hold on
plot(x,y,'k')
for x=-2:0.1:2
plot([x 0],[sqrt(2-x) sqrt(2-x)-1],'r')
plot([x 0],[sqrt(2-x) sqrt(2-x)*(1-2*x)  ],'m')
hold on
end

figure
i=1;
j=1;
for x=-2:0.1:2
    i=1;
for y=-2:0.1:2
z(i,j)=2*x^2+4*y+3;
i=i+1;
end
j=j+1;
end
 x=-2:0.1:2;
 y=-2:0.1:2;
contour(x,y,z,150)
y=-2:0.1:2;
x=2-y.^2;
hold on
plot(x,y,'k')
for x=-2:0.1:2
plot([x 0],[sqrt(2-x) sqrt(2-x)-1],'r')
plot([x 0],[sqrt(2-x) sqrt(2-x)*(1-2*x)  ],'m')
hold on
end
figure
i=1;
j=1;
for x=-2:0.1:2
    i=1;
for y=-2:0.1:2
z(i,j)=2*x^2+4*y+3;
```

```
i=i+1;
end
j=j+1;
end
 x=-2:0.1:2;
 y=-2:0.1:2;
contour(x,y,z,150)
y=-2:0.1:2;
x=2-y.^2;
hold on
plot(x,y,'k')
for x=-2:0.1:2
plot([x 0],[-sqrt(2-x) -sqrt(2-x)-1],'r')
plot([x 0],[-sqrt(2-x) -sqrt(2-x)*(1-2*x) ],'m')
hold on
end
figure
y=-2:0.1:2;
x=2-y.^2;
hold on
plot(x,y,'k')
```

Appendix B
Expectation–Maximization Algorithm

Consider the following example. The coin is randomly chosen from the two coins. Let the probability of selecting the first coin be π_1 and the probability of selecting the second coin be $\pi_2 = 1 - \pi_1$. The selected coin is tossed to obtain the outcome: head and tail. If the coin 1 is selected, the probability of obtaining head is p and the probability of obtaining tail is $1 - p$. Similarly if the coin 2 is selected, the probability of obtaining head is q and the probability of obtaining tail is $1 - q$. Given N outcomes $S = [s_1\ s_2\ \cdots s_N]$, the unknown parameters $\Theta = (\pi_1, p, q)$ are identified using Expectation–Maximization algorithm as described below. Note that s_i takes the value 0 if head is obtained and 1 if tail is obtained. It is noted that the corresponding index for the coin number is not known. The corresponding index is represented as the random vector known as latent variable vector and it is represented as $J = [J_1 J_2 J_3 \cdots J_N]$. It takes 1 if the first coin is chosen and 0 if the second coin is chosen. The unknown parameters are estimated such that $log(P(S/\Theta)) = \log(P(s_1/\Theta)P(s_1/\Theta) \cdots P(s_N/\Theta))$ is maximized. This is known as maximum likelihood estimation. The equation is rewritten as follows:

$$\log(P(S/\Theta)) = \log\left(\sum_J P(S, J/\Theta)\right)$$

where Q is the arbitrary probability. Using Gibbs' inequality, we get the following:

$$\log\left(\sum_J Q(J/S, \Theta)\frac{(P(S, J/\Theta))}{Q(J/S, \Theta)}\right) \geq \tag{B.1}$$

$$\sum_J Q(J/S, \Theta)\log\left(\frac{(P(S, J/\Theta))}{Q(J/S, \Theta)}\right)$$

The closer bound is achieved by maximizing (B.2). This is obtained as follows:

E. S. Gopi, *Digital Speech Processing Using Matlab*,
Signals and Communication Technology, DOI: 10.1007/978-81-322-1677-3,
© Springer India 2014

$$= \sum_J Q(J/S,\Theta)\log(P(J/S,\Theta) + \sum_J Q(J/S,\Theta)\log P(S/\Theta) - \sum_J Q(J/S,\Theta)\log Q(J/S,\Theta)) \tag{B.2}$$

The second term is $\log P(S/\Theta)$. Third term is the entropy H. The maximization of $\sum_J Q(J/S,\Theta)\log\left(\frac{P(J/S,\Theta)P(S/\Theta)}{Q(J/S,\Theta)}\right)$ is achieved by maximizing the first term in (B.2). Thus, the optimal probability mass function $Q(J/S,\Theta)$ is chosen as $P(J/S,\Theta)$ that maximizes (B.2). (B.2) is rewritten with $Q(J/S,\Theta) = P(J/S,\Theta)$ (without third term) as follows:

$$Q(\Theta) = \sum_J P(J/S,\Theta)\log P(J,S/\Theta) \tag{B.3}$$

$Q(\Theta)$ is viewed as the expectation of $\log P(J,S/\Theta)$ over the probability $P(J/S,\Theta)$ and is represented as $E(\log P(J,S/\Theta))$. In the example, it is computed as follows:

$$Q(\Theta) = E(\log P(J,S/\Theta)) \tag{B.4}$$

$P(J,S/\Theta)$ is computed as follows:

$$P(J_1S_1J_2S_2\cdots J_NS_N/\Theta) = P(J_1S_1/\Theta)P(J_2S_2/\Theta)\cdots P(J_NS_N/\Theta) \tag{B.5}$$

$P(J_iS_i/Q)$ takes any of the following combinations: $P(0,0/Q)$ $P(0,1/Q)$, $P(1,0/Q)$, and $P(1,1/Q)$, and it is computed as follows. $P(0,0/Q)$ is probability of obtaining head from the first coin and is equal to $P(head/coin1\ is\ selected) \times P(coin1) = p\pi_0$. Similarly, $P(1,0/Q) = q\pi_1$, $P(0,1/Q) = (1-p)\pi_0$, and $P(1,1/Q) = (1-q)\pi_1$. From the above discussion, the generalized expression to obtain $P(J_i,S_i)$ is obtained as follows:

$$P(J_i,S_i/Q) = (p^{1-S_i}(1-p)^{S_i}\pi_0)^{1-J_i}(q^{1-S_i}(1-q)^{S_i}\pi_1)^{J_i} \tag{B.6}$$

$$\Rightarrow P(J,S/\Theta) = \pi_{i=1}^{i=N}(p^{1-S_i}(1-p)^{S_i}\pi_0)^{1-J_i}(q^{1-S_i}(1-q)^{S_i}\pi_1)^{J_i} \tag{B.7}$$

$$\Rightarrow P(J,S/\Theta) = \sum_{i=1}^{i=N}(1-J_i)\log(p^{1-S_i}(1-p)^{S_i}\pi_0) + J_i\log(q^{1-S_i}(1-q)^{S_i}\pi_1) \tag{B.8}$$

$$\Rightarrow Q = E(\log(P(J,S/\Theta))) = \sum_{i=1}^{i=N}(1-E(J_i/S_i,\Theta))\log(p^{1-S_i})(1-p)^{S_i} + E(J_i/S_i,\Theta)\log(q^{1-S_i}(1-q)^{S_i}\pi_1) \tag{B.9}$$

As J_i takes the value either 0 or 1, we obtain the following. $E(J_i/S_i,\Theta) = 1\times P(J_i = 1/S_i,\Theta) + 0\times P(J_i = 0/S_i,\Theta) = P(J_i = 1/S_i,\Theta)$, which is computed as follows:

$$P(J_i = 1/S_i, \Theta) = \frac{P(S_i/J_i = 1, \Theta)P(J_i = 1/\Theta)}{P(S_i/\Theta)} \tag{B.10}$$

Note that $P(S_i/J_i = 1, \Theta) = q^{1-S_i}(1-q)^{S_i}$, $P(J_i = 1/\Theta) = \pi_1$ and $P(S_i/\Theta) = P(S_i/J_i = 0, \Theta)P(J_i = 0/\Theta) + P(S_i/J_i = 1, \Theta)P(J_i = 1/\Theta)$.
Thus

$$E(J_i/S_i, \Theta) = P(J_i = 1/S_i, \Theta) = \frac{q^{1-S_i}(1-q)^{S_i}\pi_1}{q^{1-S_i}(1-q)^{S_i}\pi_1 + p^{1-S_i}(1-p)^{S_i}\pi_0}. \tag{B.11}$$

The optimal parameters $\Theta = (p, q, \pi_0)$ are obtained by maximizing Q (B.9). This is obtained iteratively by repeating two steps.

1. Initialize the unknown parameters $\Theta = (p, q, \pi_0)$. It is represented as Θ^t with $t = 1$.
2. Compute $E(J_i/S_i, \Theta)$ using (B.9) with the latest best parameters Θ^t
3. Substituting B.11 in B.9 and differentiating with respect to p, q, and π_0 and equate to zero to obtain the unknown parameters as the next iterated best values. These are computed using (B.12–B.14).
4. Steps 2 and 3 are repeated for finite number of iterations to obtain the optimal parameters. Note that step 2 is the expectation stage and the step 3 is the maximization stage. Also, it is noted $\pi_1 = 1 - \pi_0$

$$p^{t+1} = \frac{\sum_i^{i=N} k_i S_i}{\sum_i^{i=N} k_i} \tag{B.12}$$

$$q^{t+1} = \frac{\sum_i^{i=N} (1-k_i) S_i}{\sum_i^{i=N} (1-k_i)} \tag{B.13}$$

$$\pi_0^{t+1} = \frac{1}{N}\sum_{i=1}^{i=N} k_i \tag{B.14}$$

Appendix C
Diagonalization of the Matrix

The square matrix A is said to be diagonalizable if with some arbitrary invertible matrix M, $M^{-1}AM = D$, where D is the diagonal matrix. By Schur's lemma, for any square matrix A, there exists the orthogonal matrix E such that $E^H AE = T$, where T is the triangular matrix. Note that H is the Hermitian transpose. If A is the symmetric matrix, $E^H A^H E = T^H = T \Rightarrow$ T is the diagonal matrix, and hence, we prove that the symmetric matrix is always diagonalizable with the orthonormal matrix. The steps to diagonalize the arbitrary $n \times n$ symmetric matrix A are summarized below.

1. Obtain the eigenvalues of the matrix A by solving $det(A - \lambda I) = 0$. Let it be represented as λ_i, where $i = 1, 2, 3 \cdots n$
2. Compute the eigenvectors V_i by obtaining the vectors of the null space of matrix $(A - \lambda_i I)$ corresponding to λ_i .
3. Arrange eigenvectors in the columnwise to obtain the eigen matrix E.
4. Thus, $E^H AE = D$ is obtained. D is the diagonal matrix with diagonal elements are the corresponding eigenvalues.

C.1 Positive Definite Matrix

The square matrix A is said to be positive semi-definite matrix if for some arbitrary vector X, $X^H AX \geq 0$. If the matrix A is symmetric matrix, we represent $X^H E^H AEX = (EX)^H D(EX) = Y^H DY = \lambda_1|y_1|^2 + \lambda_2|y_2|^2 + \cdots \lambda_n|y_n|^2$, where y_i is the ith element of the vector Y. Thus, if all the eigenvalues are greater than or equal to zero, we get $X^H AX \geq 0$. Hence, the matrix is said to be positive definite, if all the eigenvalues are greater than or equal to zero.

E. S. Gopi, *Digital Speech Processing Using Matlab*,
Signals and Communication Technology, DOI: 10.1007/978-81-322-1677-3,
© Springer India 2014

Appendix D
Condition Number

The linear predictive coefficients described in the Sect. 3.3 involve computation of the inverse of the symmetric matrix. The inverse of the symmetric matrix S is computed as follows. As the matrix S is always diagonalizable, it can be represented as $S = EDE^T$. The columns of the matrix E is filled up with the eigenvectors of the matrix S, and the diagonal elements of the matrix D are the corresponding eigenvalues. Thus, $S^{-1} = ED^{-1}E^T$. If the inverse of the eigenvalue is very large, it is not possible to compute the inverse matrix using computer (due to overflow). This is referred as the matrix S is not sufficiently non-singular. This is measured as the ratio of largest eigenvalue to the smallest eigenvalue. This is known as condition number of the matrix S.

D.1 Preemphasis

Speech segment is subjected to preemphasis to reduce the condition number, so that the symmetric matrix used in solving LPC becomes sufficiently non-singular. It was proved that the condition number of the symmetric matrix using in LPC is proportional to the ratio of the maximum value of the speech spectrum to the minimum value of the spectrum of the speech segment. The preemphasis is the high pass filter whose transfer function is represented as $H_{\text{preemphasis}}(Z) = 1 - \alpha z^{-1}$. The α is chosen as $1 - \frac{2f_c}{f_s}$, where f_s is the sampling frequency in Hz and f_c is usually chosen in the range between 0 to 150 Hz. The speech production model is slightly modified incorporating the preemphasis as shown in Fig. D.1. In Fig. D.1, the speech signal produced using the vocal tract is assumed as the preemphasis speech signal. That is, the output of the vocal tract is $S(Z)H_{\text{preemphasis}}(Z)$ instead of $S(Z)$. This helps to compute the LPC of the vocal tract using sufficiently non-singular matrix. To compensate the effect of preemphasis, the output of vocal tract is given as the input the filter $\frac{1}{H_{\text{preemphasis}}}$ to obtain $S(Z)$. In case of feature extraction, LPC obtained with preemphasis filter can be used. In case of speech synthesis, the effect

E. S. Gopi, *Digital Speech Processing Using Matlab*,
Signals and Communication Technology, DOI: 10.1007/978-81-322-1677-3,
© Springer India 2014

of preemphasis is compensated using the inverse filter in the output as shown in Fig. D.1. The LPC with and without pre emphasis and the corresponding spectrum are given in the figures D.2–D.5. The improvement in the condition number, achieved using the preemphasis (with $\alpha = -0.9840$), is illustrated in Fig. D.6.

```
%preemp.m
%demonstration for improving the condition number using preemphasis.
[A,FS]=wavread('D:\Gopi_academic\Dspeechprocessing\mfcc\data\train\s2.wav');
[res1,res2,speechsegment,utforste,ltforste,ltforzcr]=endpointdetection(A',FS);
S1=blkproc(speechsegment,[1 FS*30*10^(-3) ],'spe(x)');
S2=blkproc(speechsegment,[1 FS*30*10^(-3) ],'spepreemp(x)');
S3=blkproc(speechsegment,[1 FS*30*10^(-3) ],'spelpc(x)');
S4=blkproc(speechsegment,[1 FS*30*10^(-3) ],'spepreemplpc(x)');
col=[];
figure
for i=1:1:21
subplot(5,5,i)
plot(S1(1:1:188,i))
end
figure
for i=1:1:21
subplot(5,5,i)
plot(S2(1:1:188,i),'r')
end
figure
for i=1:1:21
subplot(5,5,i)
plot(abs(fft(S3(:,i),1024)))
end
figure
for i=1:1:21
subplot(5,5,i)
plot(abs(fft(S4(:,i),1024)),'r')
end
MA1=max(S1);
MI1=min(S1);
MA2=max(S2);
MI2=min(S2);
CONNUMBER1=MA1./MI1;
CONNUMBER2=MA2./MI2;
figure
stem(CONNUMBER1,'ro')
hold on
stem(CONNUMBER2,'b*')

function [res]=spe(x)
res=abs(fft(x))';

function [res]=spelpc(x)
res=lpc1(x,13)';

function [res]=spepreemp(x)
res=abs(fft(conv(x,[1 -0.9840])))';

function [res]=spepreemplpc(x)
y=conv(x,[1 -0.9840]);
res=lpc(y,13)';
```

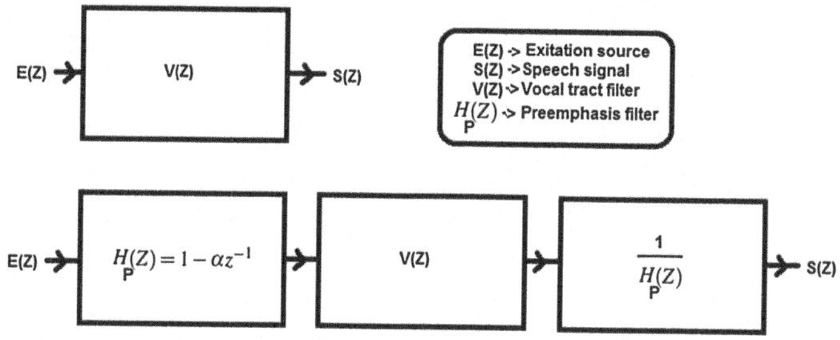

Fig. D.1 Block diagram of the speech production model, **a** without preemphasis and **b** with preemphasis

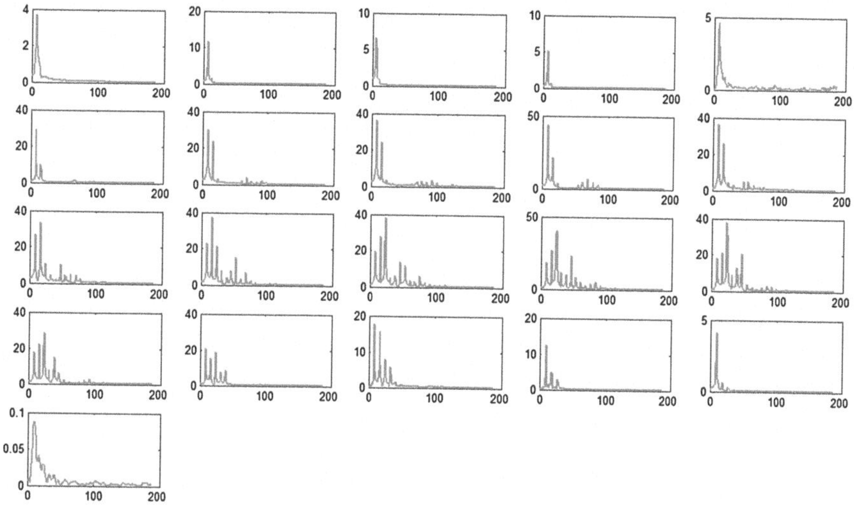

Fig. D.2 Linear predictive coefficients without preemphasis for various frames

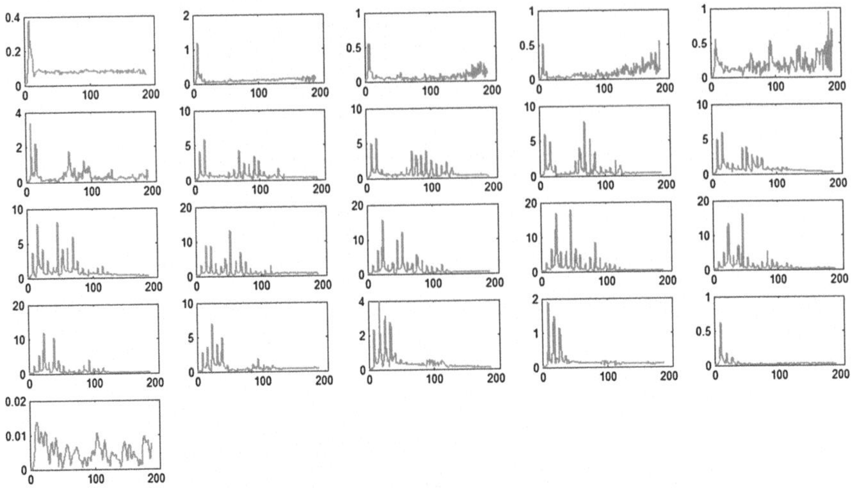

Fig. D.3 Linear predictive coefficients with preemphasis for various frames

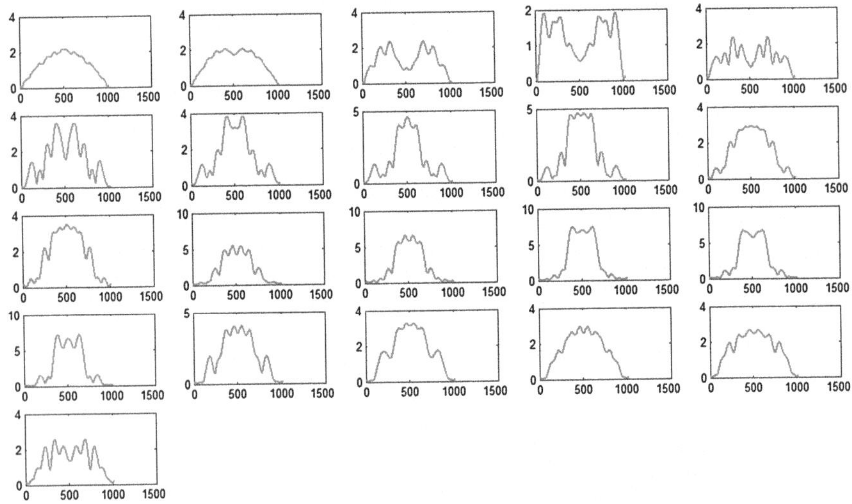

Fig. D.4 Spectrum of the speech segment without preemphasis for various frames

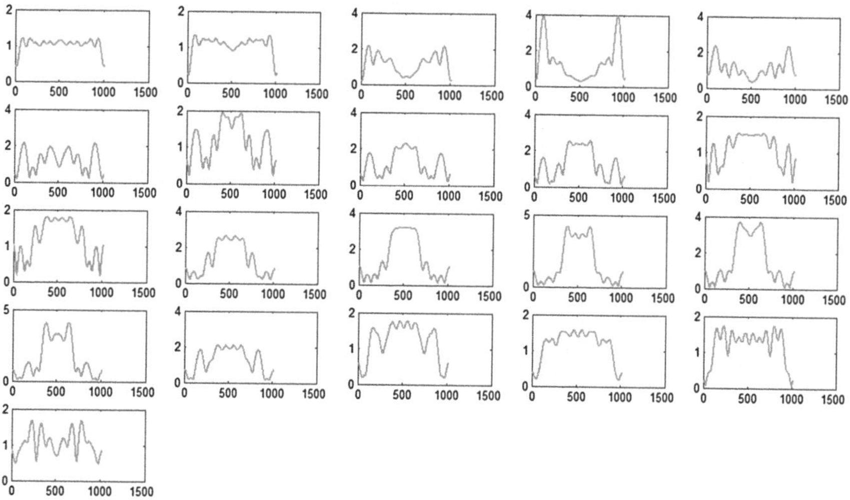

Fig. D.5 Spectrum of the speech segment with preemphasis for various frames

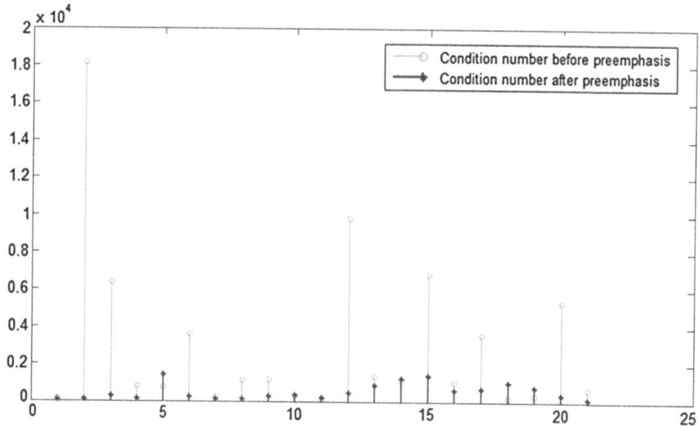

Fig. D.6 Illustration of the reduction in the condition number using the preemphasis for various frames

Appendix E
Spectral Flatness

Let $\frac{1}{L(Z)}$ be the transfer function of the vocal tract and $E(Z)$ be the Z-transformation of the excitation source. The output of the model (refer Fig. E.1) is the speech signal, and it is represented as $S(Z)$ in Z-domain. Thus, $E(Z)$ is related to $L(Z)$ and $S(Z)$ as $E(Z) = L(Z)S(Z)$. Note that $L(Z)$ is the Z-transformation of LPC filter. The order of the filter $L(Z)$ is chosen such that average power of $E(Z)$ is minimized. This can also be measured in terms of spectral flatness (white). The excitation source is assumed as the white source, and hence, its spectrum is almost flat. (Note that for the voiced speech segment, it is the addition of white source and impulse stream). Thus, it is understood that if the spectrum is flat, the model is better. This helps to fix the order of the LPC filter. The spectral flatness of the excitation source $E(Z)$ is measured by its variance. This is measured as follows.

Let the energy of the excitation source $E(Z)$ which is computed as $\frac{1}{2\pi}\int_0^{2\pi} |E(e^{jw})|^2 dw$ be represented as C. The normalized power spectrum of the excitation source is represented as $J = \frac{|E(e^{jw})|^2}{C}$. Note that the mean of the signal J is 1. Hence, the variance is computed as the average of $(J-1)^2$ over the range 0 to 2π. The function $\frac{(J-1)^2}{2}$ is approximated as $J - 1 - \log(J)$. Thus, the spectral flatness is computed as follows:

$$\mathrm{SF} = K\frac{1}{2\pi}\int_0^{2\pi} (J-1)^2 dw \tag{E.1}$$

$$\Rightarrow \mathrm{SF} = 2K\frac{1}{2\pi}\int_0^{2\pi} (J - 1 - \log(J))dw \tag{E.2}$$

where SF is the spectral flatness. K is the arbitrary constant and is chosen as $\frac{1}{2}$ for simplicity.

$$\mathrm{SF} = \frac{1}{2\pi}\int_0^{2\pi} (J - 1 - \log(J))dw \tag{E.3}$$

E. S. Gopi, *Digital Speech Processing Using Matlab*,
Signals and Communication Technology, DOI: 10.1007/978-81-322-1677-3,
© Springer India 2014

$$\Rightarrow \text{SF} = \frac{1}{2\pi}\int_0^{2\pi} J dw - 1 - \frac{1}{2\pi}\int_0^{2\pi} \log(J) dw \tag{E.4}$$

$$\Rightarrow \text{SF} = -\frac{1}{2\pi}\int_0^{2\pi} \log(J) dw \tag{E.5}$$

$$\Rightarrow \text{SF} = -\frac{1}{2\pi}\int_0^{2\pi} \log(|E(e^{jw})|^2) dw - \frac{1}{2\pi}\int_0^{2\pi} \log(C) dw \tag{E.6}$$

$$\Rightarrow \text{SF} = -\frac{1}{2\pi}\int_0^{2\pi} \log(|E(e^{jw})|) - \log(C) \tag{E.7}$$

$$\Rightarrow \text{SF} = -\frac{1}{2\pi}\int_0^{2\pi} \log(|V(e^{jw})|^2 |S(e^{jw})|^2) - \log(C) \tag{E.8}$$

$$\Rightarrow \text{SF} = -\frac{1}{2\pi}\int_0^{2\pi} \log(|V(e^{jw})|^2) - \frac{1}{\pi}\int_0^{2\pi} \log(|S(e^{jw})|^2) - \log(C) \tag{E.9}$$

Let the transfer function with single zero be represented as $H(Z) = 1 - re^{j\theta}Z^{-1}$, with the condition $0 < r < 1$. $\frac{1}{2\pi}\int_0^{2\pi} \log|H(e^{jw})|^2 dw$ is computed as follows. Let $H(e^{jw}) = re^{j\theta}$, with $r > 0$ and $\log(H(e^{jw})) = \log r + j\theta$. Hence, $\log|H(e^{jw})|$ is represented as $\log|r| = Re(\log(H(e^{jw})))$. Hence $\log|H(e^{jw})|^2$ is computed as follows:

$$\log|H(e^{jw})|^2 = 2\log|H(e^{jw})| = 2Re(\log(H(e^{jw}))) \tag{E.10}$$

$$\Rightarrow 2Re(\log(H(e^{jw}))) = 2Re(\log(1 - re^{-jw})) \tag{E.11}$$

$$2Re\left(-\sum_{n=1}^{n=\infty} \frac{(re^{j\theta}e^{-jw})^n}{n}\right) \tag{E.12}$$

$$\Rightarrow \log|H(e^{jw})|^2 = -\sum_{n=1}^{n=\infty} \frac{r^n cos(\theta - w)n}{n} \tag{E.13}$$

Using (E.13), we observe, $\frac{1}{2\pi}\int_0^{2\pi} \log|H(e^{jw})|^2 dw = 0$. Consider the transfer function with two zeros as $H(Z) = (1 - r_1 e^{j\theta_1} Z^{-1})(1 - r_2 e^{j\theta_2} Z^{-1}) = H_1(Z)H_2(Z)$ with the condition $0 < r_1, r_2 < 1$. $\log|H(e^{jw})|^2 = \log|H_1(e^{jw})|^2 + \log|H_2(e^{jw})|^2$. From above, we get

$$\frac{1}{2\pi}\int_0^{2\pi} \log|H(e^{jw})|^2 dw = \frac{1}{2\pi}\int_0^{2\pi} \log|H_1(e^{jw})|^2 dw + \frac{1}{2\pi}\int_0^{2\pi} \log|H_2(e^{jw})|^2 dw = 0 \tag{E.14}$$

Hence, if all the zeros of the filter lies within the unit circle, $\frac{1}{2\pi}\int_0^{2\pi} \log|H(e^{jw})|^2 dw$. As all the poles of the vocal tract filter, equivalently all the zeros of the transfer function $V(Z)$ lies within the unit circle, the first term in (E.9) is zero, and hence, SF is computed as follows:

$$SF = -\frac{1}{2\pi}\int_0^{2\pi} \log(|S(e^{jw})|^2) - \log(C) \tag{E.15}$$

E.1 Demonstration on Spectral Flatness

For the isolated speech segment, LPC is obtained with the order ranging from 9 to 150 and the corresponding spectral flatness is computed using (E.10) for 21 frames. Figure E.1 shows that SF decreases with the increasing order of the LPC. The power spectrum of the exited source is plotted for the order 9 (refer Fig. E.2) and 150 (refer Fig. E.3). It is seen that the spectrum is almost flat for the order 150.

```
%spectralflatnessdemo.m
[A,FS]=wavread('D:\Gopi_academic\Dspeechprocessing\mfcc\data...
\train\s2.wav');
[res1,res2,speechsegment,utforste,ltforste,ltforzcr]=...
endpointdetection(A',FS);
for i=9:1:150
S{i}=blkproc(speechsegment,[1 FS*30*10^(-3) ],'spectralflatness(x,P1)',i);
end
for i=9:1:150
E{i}=blkproc(speechsegment,[1 FS*30*10^(-3) ],...
'spectralflatnesserror(x,P1)',i);
end
figure
for i=9:1:9
    for j=1:1:21
        subplot(5,5,j)
        plot(E{i}(:,j))
    end
end
figure
for i=150:1:150
    for j=1:1:21
        subplot(5,5,j)
        plot(E{i}(:,j))
    end
end
S1=cell2mat(S');
figure
imagesc(S1)

%spectralflatness.m
function [res]=spectralflatness(x,n)
res=lpc(x,n)';
e=abs(fft(res,1024)).*abs(fft(x,1024))';
e1={log}(sum(e.^2)/1024);
s=(abs(fft(x,1024)).^2);
s1=sum(log(s))/1024;
res=e1-s1;
```

```
%spectralflatnesserror.m
function [e]=spectralflatnesserror(x,n)
res=lpc(x,n)';
e=abs(fft(res,1024)).*abs(fft(x,1024))';
```

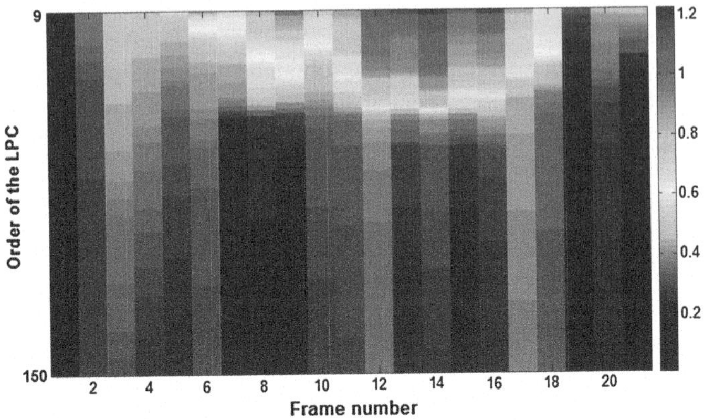

Fig. E.1 Illustration of the reduction in the SF with increase in the order of the LPC filter

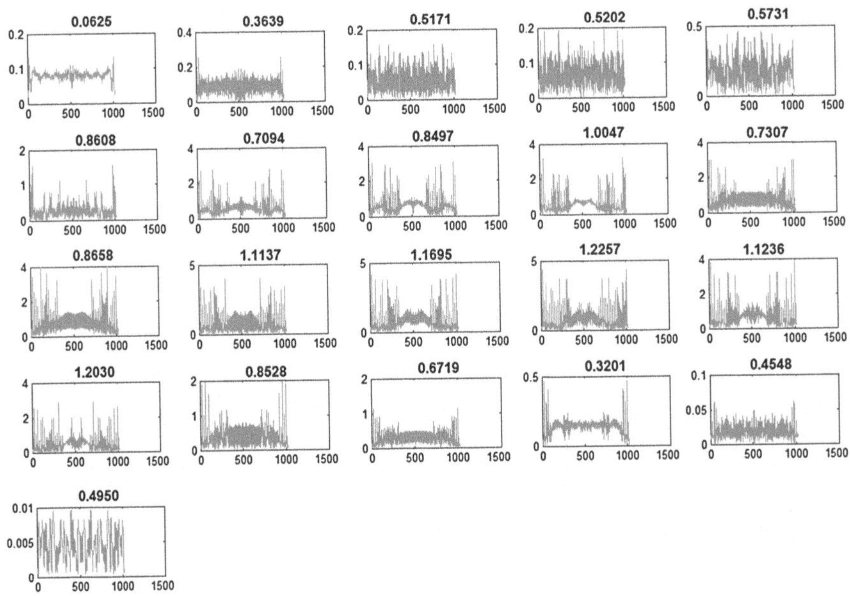

Fig. E.2 Power spectrum of the excitation source obtained as the product of $V(Z)S(Z)$ with ninth-order LPC for 21 frames. The spectrum is plotted for range 0–2π

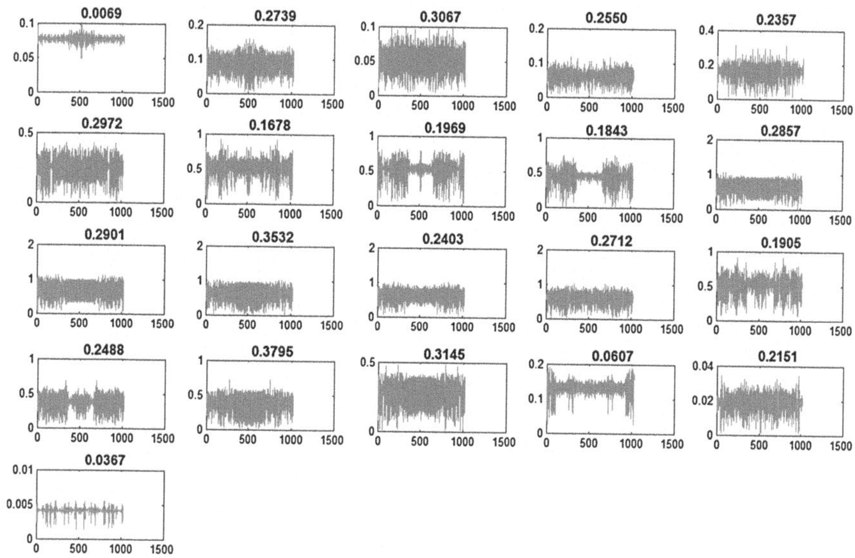

Fig. E.3 Power spectrum of the excitation source obtained as the product of $V(Z)S(Z)$ with 150th order LPC for 21 frames. The spectrum is almost flat in this case. The spectrum is plotted for range 0–2π

Appendix F
Functional Blocks of the Vocal Tract and the Ear

F.1 Vocal Tract Model

The air coming out of the lungs is mixed with the sound wave produced using the vocal chord vibration and are passed through the pharynx cavity. The air coming out of the pharynx cavity is passed through the nasal cavity and the mouth cavity. Velum acts as the switch that controls the flow of air passing through the nasal cavity. Thus, the air coming out of the nose and the mouth forms the speech signal. If the excitation source consists of air coming out of the lungs and the vocal chord vibration, the corresponding speech signal is known as voiced speech signal. If the excitation source consists of only the air coming out of the lungs, it is known as unvoiced speech signal. The functional block of vocal tract is given in Fig F.1

F.2 Ear Model

Human ear is divided into three major parts: inner ear, middle ear, and the outer ear. The outer ear consists of pinna and the external auditory canal. This helps to collect the sound energy from the outside source to the inner ear. Middle ear consists of ear drum and auditory ossicles. Sound wave is passed through the auditory canal and vibrates the ear drum. This helps to reproduce the sound wave. Auditory ossicles consist of three parts, namely malleus, incus, and stapes. They are the bones connected one after the other. The malleus is connected to the ear drum. This helps to carry the sound wave to the inner ear. The stapes are connected to oval window. Oval window helps to carry the sound wave to cochlea . Cochlea is the tube-like structure filled with fluid. The sound wave disturbs the fluid in the cochlea through oval window. The center of the cochlea consists of thousands of microscopic hairs. The different parts of the cochlear responds to different range of frequencies of the sound wave and the corresponding microscopic hairs are vibrated. The vibrations of the microscopic

E. S. Gopi, *Digital Speech Processing Using Matlab*,
Signals and Communication Technology, DOI: 10.1007/978-81-322-1677-3,
© Springer India 2014

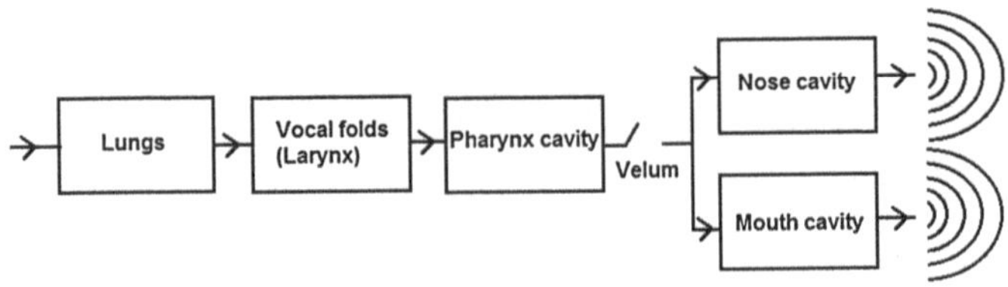

Fig. F.1 Functional block diagram of the vocal tract

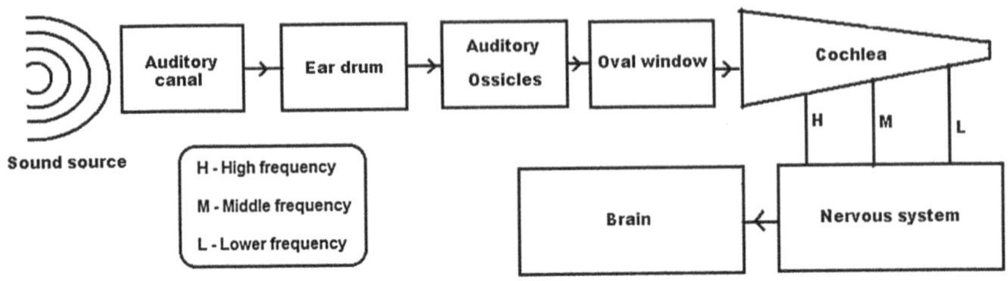

Fig. F.2 Functional block diagram of the human ear

hair send the signals to the brain to recognize the sound wave. The basic functional blocks of the human ear are given in Fig. (F.2).

About the Author

E. S. Gopi has authored four books, of which three have been published by Springer. He has also contributed 1 book chapter to book published by Springer. He has several papers in international journals and conferences to his credit. He has 15 years of teaching and research experience. He is currently Assistant Professor, Department of Electronics and Communication Engineering, National Institute of Technology, Trichy, India. He has 71 citations with h-index 4 (based on Google Scholar). His books are widely used all over the world. His research interests include pattern recognition, digital signal processing, and biologically inspired algorithms. The India International Friendship Society (IIFS) has awarded him the "Shiksha Rattan Puraskar Award" for his meritorious services in the field of education. The award was presented by Dr. Bhishma Narain Singh, former Governor, Assam and Tamil Nadu, India. He was also awarded with the "Glory of India Gold Medal" by International Institute of Success Awareness. This award was presented by Shri Syed Sibtey Razi, former Governor of Jharkhand, India. He was also awarded with "Best Citizens of India 2013" by International Publishing House.

E. S. Gopi, *Digital Speech Processing Using Matlab*,
Signals and Communication Technology, DOI: 10.1007/978-81-322-1677-3,
© Springer India 2014

About the Book

Digital Speech Processing Using Matlab deals with digital speech pattern recognition, speech production model, speech feature extraction, and speech compression. The book is written in a manner that is suitable for beginners pursuing basic research in digital speech processing. Matlab illustrations are provided for most topics to enable better understanding of concepts. This book also deals with the basic pattern recognition techniques (illustrated with speech signals using Matlab) such as PCA, LDA, ICA, SVM, HMM, GMM, BPN, and KSOM.

E. S. Gopi, *Digital Speech Processing Using Matlab*,
Signals and Communication Technology, DOI: 10.1007/978-81-322-1677-3,
© Springer India 2014

Index

E. S. Gopi, *Digital Speech Processing Using Matlab*,
Signals and Communication Technology, DOI: 10.1007/978-81-322-1677-3,
© Springer India 2014

The manufacturer's authorised representative in the EU is Springer Nature Customer Service Centre GmbH, Europaplatz 3, 69115 Heidelberg, Germany. If you have any concerns regarding our products, please contact ProductSafety@springernature.com

Printed and bound by CPI Group (UK) Ltd, Croydon, CR0 4YY
22/09/2026
02220982-0002